AF334671

Astronomy

The Alexandrian astronomer Ptolemy (fl AD 150) using a quadrant for determining the altitude of a celestial body. Behind him stands the muse of astronomy, and at the lower left of the picture is an armillary sphere, a popular device in the fifteenth century and later, on which parallels of latitude and other fundamental celestial and terrestrial circles were represented by metal rings. In this picture the artist has made the mistake, common in the Middle Ages and in the Renaissance, of confusing Ptolemy the astronomer with the Ptolemies who ruled Egypt and established the famous Library and Museum at Alexandria in the fourth century BC; this is why the astronomer is shown wearing a crown. From Gregor Reisch, Margarita Philosophica *(The Philosophical Pearl), Basle, 1508.*

ILLUSTRATED SOURCES IN HISTORY

Astronomy

COLIN A. RONAN

DAVID & CHARLES: Newton Abbot

BARNES & NOBLE BOOKS: New York
(a division of Harper & Row Publishers, Inc.)

This edition first published in 1973
in Great Britain by
David & Charles (Publishers) Limited
Newton Abbot   Devon
in the U.S.A. by
Harper & Row Publishers Inc.
Barnes & Noble Import Division
0 7153 5593 7 (*Great Britain*)
06-495970-8 (*United States*)

Set in 10 on 12pt Plantin
by C. E. Dawkins (Typesetters) Limited
London SE1
and printed in Great Britain
by Clarke Doble & Brendon Limited
Plymouth

Contents

Preface

At the present time, astronomical ideas and theories of the universe are being dramatically changed by new observing techniques such as radio telescopes and space probes. This would seem, then, to be a good moment, when everything is in the melting pot, to look back over the past two millenia or so and trace the struggles, examine the highlights and even probe some of the dead ends of astronomy. The extracts have been chosen with this in mind; to show how the concept of the universe has developed from that of a comparatively small geocentric sphere, to a boundless space of almost unimaginable proportions. Scientific books, articles and, in some cases, photographs and drawings have all been explored to provide the original source material.

All the same, the selection of original material has been mainly a matter of deciding—often regretfully—what to omit rather than of finding suitable extracts to put in. Yet it is hoped that the extracts that have been selected will help in showing the development of what seems clearly to be the earliest subject to become a true science.

In the extracts, explanatory comments are given in square brackets [ ]. Every extract has been given its original spelling, and the source for each is given in a separate list. Sources for the illustrations are acknowledged at the end of the book. The connecting text has been kept as brief as possible, but at the same time I hope that it gives enough to help the reader to appreciate the extracts.

Colin A. Ronan,
Cowlinge,
Suffolk.
1971 May 7

CHAPTER ONE

The Planetary System

To early astronomers, the heavens were an expression of divine order. The inexorable rising and setting of the Sun and Moon, the majestic pageant of the night sky with its unchanging patterns of stars, displayed the eternal, immutable nature of the gods. It was an inspired conception, magnificent in the breadth of its vision, but marred by two serious flaws—the manifestation of strange and apparently capricious events in the heavens that seemed incongruous in any divinely ordered universe, and the persistent habit of a few bright stars of wandering to and fro across the background formed by the others. If a truly heavenly cosmos surrounded the Earth, then some explanation had to be found for these imperfections.

Unexpected happenings in the sky such as haloes round the Moon, the unheralded appearance of stars with beards and long hairy tails (comets), and the sudden advent of shooting stars which seemed to drop out of the heavens without warning, could all be accounted for by classifying them as meteorological.

They were therefore regarded as effects occurring in the space surrounding the Earth, and not far off in the sky. Thus the heavens could remain inviolate from sudden disturbance, even from change and decay, since only in the upper air did transitory events appear and, by common consent, it became accepted that at their greatest height they never reached as far as the Moon, the nearest of the truly heavenly bodies.

But the wandering stars could not be explained so easily: no classification could remove them from the sky and bring them earthwards and, as it turned out, the central problem of all early astronomy was to find an explanation that would account for the observed motions, and yet do no violence to the concept of a divine cosmos.

The motions to be accounted for were complex. The Sun and Moon appeared to be the most straightforward, since they obviously moved in circles around the Earth, although their motions were faster in winter than in summer. Moreover, they were not always the same size, although it must be admitted that the variation was small and would escape all but the most experienced and careful observers. The Morning and Evening Stars —the planets Mercury and Venus—both appeared either at dawn or dusk and always kept close to the Sun, but not so the other three planets, Mars, Jupiter and Saturn. These might be found anywhere in the sky along the Sun's path. And this was not all, for the planets moved backwards as well as forwards. Certainly Saturn, Jupiter and Mars seemed to pursue a mainly eastward course in the sky; but from time to time they would stop, retrace their steps, stop, and then move on again. Mercury and Venus were a law unto themselves, and perpetually moved from one side of the Sun to the other. Here, then, was a complex of motions to be embodied into one scheme, and it is small wonder that astronomers were occupied trying different explanations until the seventeenth century AD when, at last, a general solution to the problem was evolved.

It is to the Greeks that we must turn for the first comprehensive theory, a theory that was constructed by a number of scientifically minded men over a long period, beginning in the latter part of the sixth century BC with Pythagoras, who claimed that the planets and the sphere of stars rotated about the Earth, which he thought of as a sphere fixed in the centre of the cosmos. The planets, he said, moved in circles at a regular rate, for this was the only kind of motion aesthetically acceptable for celestial bodies. But since observation did not allow so straightforward an explanation, Pythagoras proposed using a series of regular circular motions to account for planetary movements and, in essence, this is what astronomers continued to do for the next two thousand years. Yet some of Pythagoras' disciples (for he was a religious as well as a philosophical leader) modified his ideas because they believed that the universe must contain ten bodies. Consequently they imagined that there was a central fire in the midst of all creation and, around it, orbiting in perfect circles, went the Sun, the Moon, the five planets and, with them,

the Earth and another body called the 'counter-Earth'— thus obtaining their total of ten. The counter-Earth always lay between the central fire and the Earth itself, thus effectively preventing anyone observing whether or not the central fire existed. But the theory had few astronomical advantages, and soon lost favour.

The writings of Pythagoras or of Philolaos, who was the chief protagonist of the central fire theory, have not come down to us. Such as we know, we know from what other philosophers have recorded. But as it was a fixed Earth set in the centre of the universe that Greek astronomers finally settled for, this is the hypothesis that we must examine.

1 Aristotle on whether or not the Earth is fixed in space

The fact that the Earth could not move and must be absolutely still had many advocates, but none more eloquent and influential than Aristotle (384-322 BC) Born in the city of Stagiros, at seventeen he went to Athens to study under Plato at the Academy, and remained in the city for the next twenty years. In due course he became tutor to the young Macedonian prince, Alexander (Alexander the Great), but in 337 BC he returned to Athens where he established his own school— the Lyceum. Here he studied, taught, and wrote on philosophical and metaphysical problems, on politics and ethics, biological subjects, and the whole gamut of physics and astronomy: in short, he was a polymath of extraordinary breadth, even in a day when specialisation was unknown. In his astronomical book, known now by the title of the Latin translation DE CAELO *(On the Heavens), Aristotle examined the question of whether or not the Earth moves, and provided strong arguments for*

EMPIREVM
HABITACVLVM
COELVM
DEI
ELECTORVM
ET OMNIVM
Decimũm Coelũm
Primũ Mobile
Nonũ Coelium
Cristallinũm
Octauũm
Firmamentũ
COELV SATVRNI
IOVIS
MARTIS
SOLIS
VENERIS
MERCVRII
LVNÆ

Let us first decide the question whether the Earth moves or is at rest. For, as we said, there are some who make it one of the stars, and others who, setting it at the centre, suppose it to be 'rolled' and in motion about the pole as axis. That both views are untenable will be clear if we take as our starting-point the fact that the Earth's motion, whether the Earth be at the centre or away from it, must needs be a constrained motion. It cannot be the motion of the Earth itself. If it were, any portion of it would have this movement; but in fact every part moves in a straight line to the centre. Being, then, constrained and unnatural, the movement could not be eternal. But the order of the universe is eternal. Again, everything that moves with the circular movement, except the first sphere, is observed to be passed, and to move with more than one motion. The Earth then, also, whether it move about the centre or as stationary at it, must necessarily move with two motions. But if this were so, there would have to be passings and turnings of the fixed stars. Yet no such thing is observed. The same stars always rise and set in the same parts of the Earth.

Further, the natural movement of the Earth, part and whole alike, is to the centre of the whole [universe]—whence the fact that it is now actually situated at the centre—but it might be questioned, since both centres are the same, which centre it is that portions of the Earth and other heavy things move to. Is this their goal because it is the centre of the Earth or because it is the centre of the whole [universe]? The goal, surely, must be the centre of the whole. For fire and other light things move to the extremity of the area which contains the centre. It happens, however, that the centre of the Earth and of the whole [universe] is the same. Thus they do move to the centre of the Earth, but accidentally, in virtue of the fact that the Earth's centre lies at the centre of the whole [universe]. That the centre of the

Earth is the goal of their movement is indicated by the fact that heavy bodies moving towards the Earth do not move parallel but so as to make equal angles, and thus to a single centre, that of the Earth. It is clear, then, that the Earth must be at the centre and immovable, not only for the reasons already given, but also because heavy bodies forcibly thrown quite straight upward return to the point from which they started, even if they are thrown to an infinite distance. From these considerations then it is clear that the Earth does not move and does not lie elsewhere than at the centre of the universe.

With the Earth fixed at the centre of the cosmos, the task of the theoretical astronomer was to choose a combination of regular circular motions to account for the movements of the planets and of the Sun and the Moon. In Aristotle's day a complex system of concentric spheres was used, but in time this gave way to the geometrical device of the deferent and epicycle—a large circle with a small one rotating on its circumference— which permitted mathematicians to adjust arrangements until observations were accounted for with a high degree of precision.

2 Ptolemy on the spherical motions of the Heavens

The fully developed Greek planetary theory was enshrined in one of the most famous books in the history of science, the ALMAGEST, *for it was from this book that later western civilisation was to learn its astronomy. The author was Claudius Ptolemaeus, who was born in the city of Ptolemaïs Hermii on the Nile, but spent his working life at the great Library and Museum at Alexandria. Although the dates of Ptolemy's birth and death are unknown, from internal evidence it is apparent that the book was written between 127 and 151 AD. It is a veritable encyclopaedia of Greek astronomy, in which Ptolemy disposes of a moving Earth (since this had been proposed again since Aristotle's time), and demonstrates how the heavens move:*

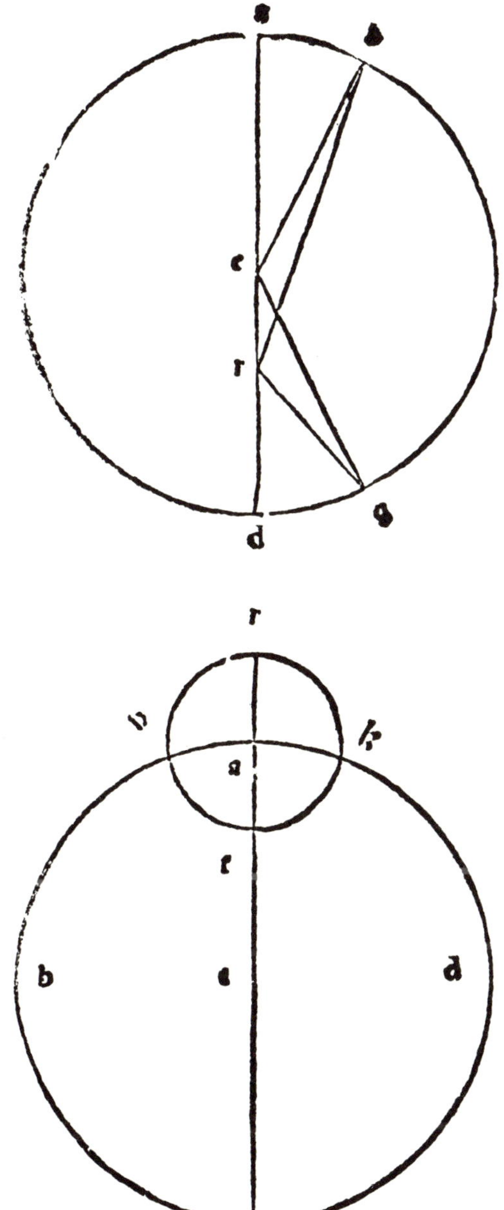

It is probable the first notions of these things came to the ancients from some such observation as this. For they kept seeing the Sun and Moon and other stars always moving from rising to setting in parallel circles, beginning to move upward from below as if out of the Earth itself, rising little by little to the top, and then coming around again and going down in the same way until at last they would disappear as if falling into the Earth. And then again they would see them, after remaining some time invisible, rising and setting as if from another beginning; and they saw that the times and also the places of rising and setting generally corresponded in an ordered and regular way.

But most of all the observed circular orbit of those stars which are always visible [the 'circumpolar' stars], and their revolution about one and the same centre, led them to this spherical notion.

3 Ptolemy on the hypothesis of regular circular movement

But the spherical motion of the planets is not regular; even the Sun does not move at a regular rate across the sky. Ptolemy was well aware of this and discusses it:

Since the next thing is to explain the apparent irregularity of the Sun, it is first necessary to assume in general that the motions of the planets in the direction contrary to the movement of the heavens are all regular and circular by nature, like the movement of the universe in the other direction. That is, the straight lines, conceived as revolving the stars or their circles, cut off in equal times on absolutely all circumferences equal angles at the centres of each; and their apparent irregularities result from the positions and arrangements of the circles on their spheres through which they produce these movements, but no departure from their unchangeableness

2 *A diagram from a very early printing of Ptolemy's* Almagest, *showing the eccentric circle, and the epicycle and deferent. (See extract 3.) This edition was published in 1515 at Venice*

has really occurred in their nature in regard to the supposed disorder of their appearances.

But the cause of this irregular appearance can be accounted for by as many as two primary simple hypotheses. For if their movement is considered with respect to a circle in the plane of the ecliptic [apparent path of the Sun in the sky] concentric with the cosmos so that our eye is the centre, then it is necessary to suppose that they make their regular movements either along circles not concentric with the cosmos, or along concentric circles [deferents]; not with these simply, but with other circles borne upon them called epicycles. For according to either hypothesis it will appear possible for the planets seemingly to pass in equal periods of time, through unequal arcs of the ecliptic circle which is concentric with the cosmos [see figure 2].

Ptolemy adopted the deferent and epicycle to account for the observed irregularity of planetary motion, and his became the standard procedure for the next 1,400 years.

4 Copernicus on the fixity of the Earth

In 1543 at Nuremberg, there appeared a book with the title DE REVOLUTIONIBUS ORBIUM COELESTIUM (On the Revolutions of the Heavenly Spheres). *Written by Nicholas Copernicus (1473-1543), a canon of the Roman Catholic cathedral at Frauenberg on the shores of the Gulf of Danzig (Gdansk), this publication became the focal point for what was to be nothing less than a complete overthrow of Ptolemaic astronomy. Copernicus argued that Ptolemy and others had been wrong in thinking that violent motions would occur if the Earth moved:*

. . . they say that the Earth remains at rest at the middle of the world [universe] and that there is no doubt about this. But if someone opines that the Earth revolves, he will also say that the movement is natural and not violent. Now things which are according to nature produce effects contrary to those which are violent. For things to which force or violence is applied get broken up and are unable to subsist for a long time. But things which are caused by nature are in a right condition and are kept in their best organisation. Therefore Ptolemy had no reason to fear that the Earth and all things on the Earth would be scattered in a revolution caused by the efficacy of nature, which is greatly different from that of art or from that which can result from the genius of man. But why did he not feel anxiety about the world instead, whose movement must necessarily be of greater velocity, the greater the heavens are than the Earth? Or have the heavens become so immense, because an unspeakably vehement motion has pulled them away from the centre, and because the heavens would fall if they came to rest anywhere else?

Surely if this reasoning were tenable, the magnitude of the heavens would extend infinitely. . . .

But let us leave to the philosophers of nature the dispute as to whether the world is finite or infinite, and let us hold as certain that the Earth is held together between its two poles and terminates in a spherical surface. Why therefore should we hesitate any longer to grant to it the movement which accords naturally with its form, rather than put the whole world in a commotion—the world whose limits we do not and cannot know? And why not admit that the appearance of daily revolution belongs to the heavens but the reality belongs to the Earth? And things are as when Aeneas said in Virgil: 'We sail out of the harbour, and the land and the cities move away'. As a matter of fact, when a ship floats on over a tranquil sea, all the things outside seem to the voyagers to be moving in a movement which is the image of their own, and they think on the contrary that they themselves and all the things with them are

3 *The solar system as envisaged by Copernicus, and published in his* De Revolutionibus Orbium Coelestium *(Nuremberg 1543). This diagram follows the section quoted in extract 5*

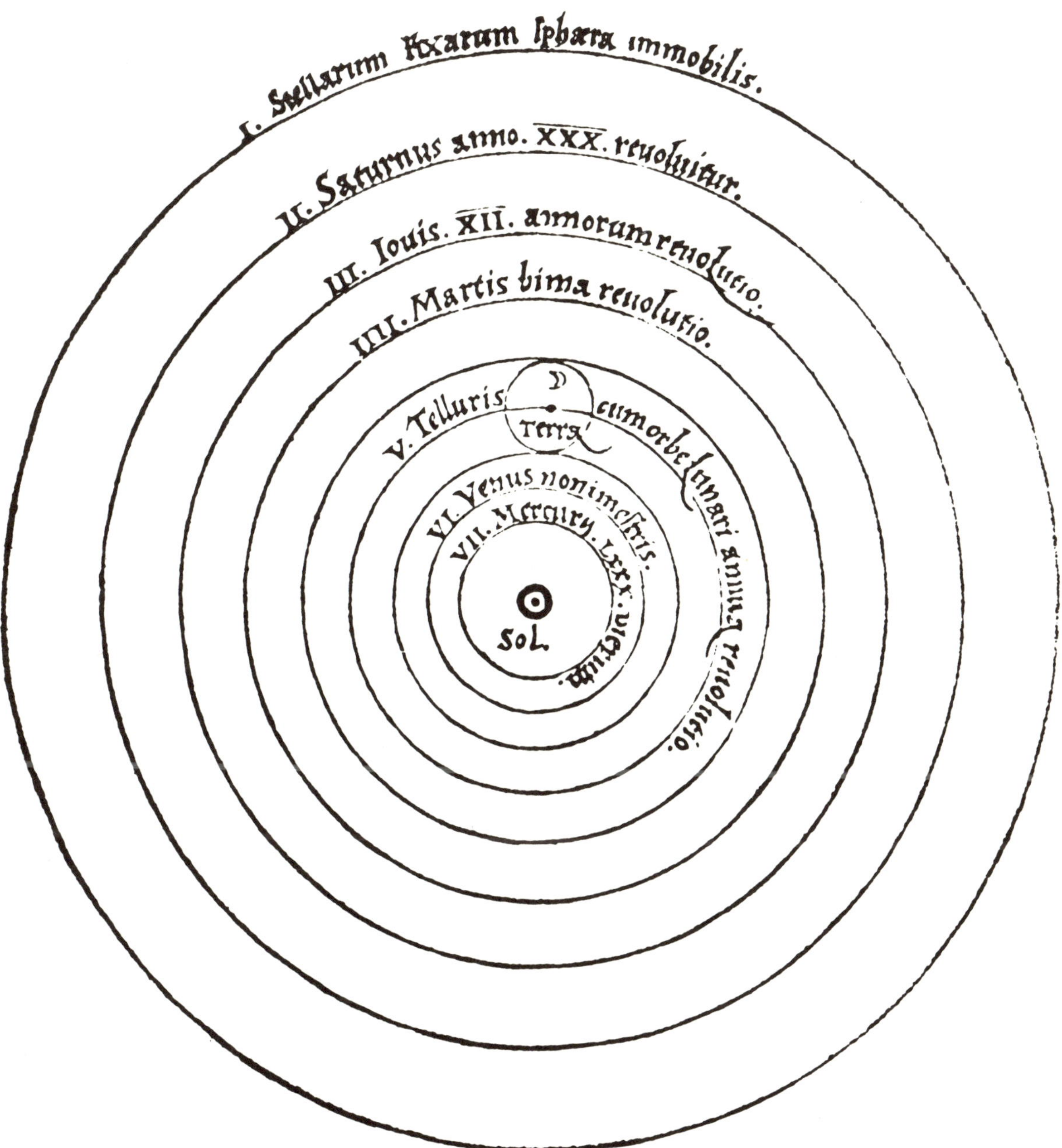

I. Stellarum fixarum sphæra immobilis.
II. Saturnus anno. XXX. reuoluitur.
III. Iouis. XII. annorum reuolutio.
IIII. Martis bima reuolutio.
V. Telluris cum orbe lunari annua reuolutio.
Terra
VI. Venus nonimestris.
VII. Mercury. LXXX. dierum.
SOL.

orbit in circular paths at a regular unvarying velocity. He discovered this first for the orbit of Mars, but it was only after many years of tedious computation that he was able to prove that what was true for Mars was true also for the other planets. Between 1618 and 1621 he published an explanation for the general reader in his

EPITOME ASTRONOMIAE COPERNICAE (*Epitome of Copernican Astronomy*), *which he cast in the then popular form of questions and answers:*

If you set up no solid spheres in the heavens and if all the movements of the planets are regulated by natural faculties, which are implanted in the bodies of the planets: then I ask what will the theory of astronomy be? For it seems that the theory cannot do without the imagining of circles and spheres.

It can easily do without the useless furniture of fictitious circles and spheres. But there is such great

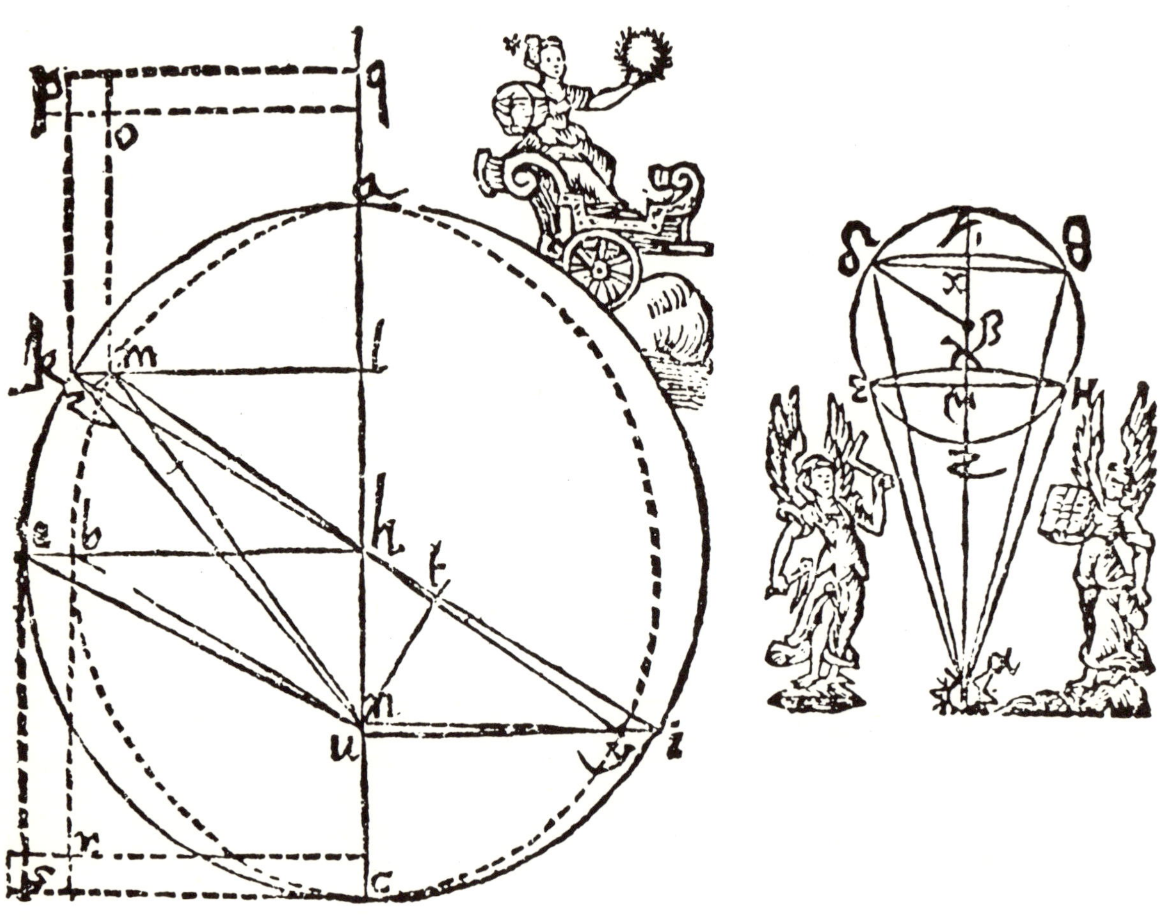

5 The eccentric orbit of Mars as illustrated in Kepler's Astronomia Nova (*Heidelberg 1609*) *of which only comparatively few copies were printed. It was in this that he announced his preliminary results on elliptical orbits obtained from his study of Tycho Brahe's observations*

need of imagining the true figures, in which the routes of the planets are arranged, that we are impoverishing Astronomy and that the big job to be worked on by the true astronomer is to demonstrate from observations what figures the planetary orbits possess: and to devise such hypotheses, or physical principles as can be used to demonstrate the figures which are in accord with the deductions made from observations.

7 Kepler's first and second laws of planetary motion

Kepler's 'physical principles' were based on his theory of a magnetic force emanating from the Sun. He had read with great interest De Magnete, *William Gilbert's book on magnetism, published in London in* 1600, *and he conceived the idea of magnetic 'threads' extending from the Sun to the planets. These he discussed earlier in the book, and now he used them to provide a physical basis for his novel ideas of planetary orbits:*

The what sort of figure of the planetary orbit is formed according to the physical principles of Book IV?

. . . because we have laid down that in the body of any planet there are present twofold threads: therefore by the mingling of the faculties of the planet's body and the Sun's motor power, (1) the planet describes an orbit oblique to the ecliptic; and because the threads of latitude remain in approximately a parallel posture during the whole circuit but not wholly so, and hence they are deflected gradually after many revolutions: therefore (2) the plane contained by the orbit of the planet is approximately a perfect plane but not wholly so; and, hence, when one revolution has been completed, the centre of the planetary globe does not return exactly to its starting point, but it entwines a new circle with the first circle traversed and described—after the fashion of the circles of the . . . thread which the silk-worm drops, throwing the thread around itself and building a little house by the interweaving of many entwined circles.

For this reason the longest digressions in latitude are not found in all ages in the same parts of the zodiac. And because the threads of libration make the planet to be drawn away from the Sun to one side but then to be driven away from that region; accordingly the planet (3) describes its orbit around the Sun but not as around its own centres, that is to say, it describes an orbit eccentric to the Sun: and for this reason the orbit is not (4) a perfect circle, but one slightly narrower and more pressed in on the sides, like the figure of an ellipse. (5) For the same cause and because the form of the solar body, which form gives movement to the planet, is slighter and weaker in the larger circle, it is not possible for the planet to be moved with the same speed in all parts of its orbit; but the planet is slow at a long distance from the Sun and fast at a short distance.

Here we have Kepler propounding what came to be known as his first law of planetary motion (motion in an ellipse), and the basis of his second law (that a planet moves faster when near the Sun and slower when further off), a law which he explained in complex detail later on. But Kepler is also credited with a third law of planetary motion, and this law came earlier in the book, in his ideas of the power emanating from the Sun:

8 Kepler's third law of planetary motion

By what reasons are you led to make the Sun the moving cause or the source of movement for the planets?

1. Because it is apparent that in so far as any planet is more distant from the Sun than the rest, it moves more slowly—so that the ratio of the periodic times is the ratio of 3/2th powers of the distances from the Sun. Therefore we reason that the Sun is the source of movement.

Although the first and third laws became generally accepted, it took time before the second law was agreed in all its geometrical complexity. And Kepler's physical principles were never widely adopted, for it was soon realised at this time that the whole question of the motion

6 *Kepler believed his greatest achievement was his discovery of musical harmonies for the planets: these he believed were the essence of the divine harmony of the universe. He gave his results in his* Harmonices Mundi *(Linz 1619) and the page shown here gives the 'divine' musical scales for the planets, calculated from the velocities of the planets when both closest and furthest from the Sun in their elliptical paths*

of bodies in space and on Earth needed radically re-thinking.

9 Galileo on motion

This was to a great extent achieved by Galileo Galilei (1564-1642) who, while under house arrest near Florence for his open support of the Copernican theory, wrote his DISCORSI . . . À DUE NUOVE SCIENZE *(Discourses . . . on two new sciences). The book was smuggled out of Italy and published in Leyden in 1638, and contains the important principle that a force does not need to be applied to a body all the time to keep it moving—a mistaken view that Kepler still held:*

Furthermore we may remark that any velocity once imparted to a moving body will be rigidly maintained as long as the external causes of acceleration and retardation are removed, a condition which is found only on horizontal planes; for in the case of planes which slope downwards there is already present a cause of acceleration, while on planes sloping upward there is retardation; from this it follows that motion along a horizontal plane is perpetual: for, if the velocity be uniform, it cannot be diminished or slackened, much less destroyed. Further, although any velocity which a body may have acquired through natural fall is permanently maintained so far as its own nature is concerned, yet it must be remembered that if, after descent along a plane inclined downwards, the body is deflected to a plane inclined upward, there is already existing in this latter plane a cause of retardation; for in any such plane this same body is subject to a natural acceleration downwards. Accordingly we have here the superposition of two different states, namely, the velocity acquired during the preceding fall which if acting alone would carry

7 *In 1630 Galileo's* Dialogo dei Massimi Sistemi *(Dialogue On The Great World Systems) was published at Florence. Written, as its title states, in the form of a dialogue, it was a discussion on the relative merits of the Aristotelian, Ptolemaic and Copernican systems of the universe—about the old geocentric and the new heliocentric hypotheses, in fact. This beautiful frontispiece of the first edition shows (from left to right), apocryphal portraits of Aristotle, Ptolemy and Copernicus*

20

DIALOGO
di
GALILEO GALILEI LINCEO
AL SER.mo FERD. II. GRAN. DVCA DI
TOSCANA

que talis pofitio. Media Stella oriétali quam pro-
Ori.   ** ✶   *   Occ.

xima min. tantum fec. 20. elongabatur ab illa, &
a linea recta per extremas, & Iouem producta
paululum verfus auftrum declinabat.

   Die 18. hora 0. min. 20. ab occafu, talis fuit a-
fpectus. Erat Stella orientalis maior occidenta-
Ori.   *   ✶   *   Occ.

li, & a Ioue diftans min. pr. 8. Occidentalis vero
a Ioue aberat min. 10.

   Die 19. hora noctis fecunda talis fuit Stellarũ
coordinatio: erant nempe fecundum rectam li-
Ori.   *   ✶   *   *   Occ.

neam ad vnguem tres cum Ioue Stellæ: Orientalis vna a Ioue diftans min. pr. 6. inter Iouem, &
primam fequenté occidentalem, mediabat min.
5. interftitium: hæc autem ab occidentaliori a-
berat min. 4. Anceps eram tunc, nunquid inter
orientalem Stellam, & Iouem Stellula mediaret,
verum Ioui quam proxima, adeo vt illum fere
tangeret; At hora quinta hanc manifefte vidi
medium iam inter Iouem, & orientalem Stellam
locum exquifite occupantem, ita vt talis fuerit
Ori.   * * ✶   * *   Occ.

configuratio. Stella infuper nouiffime confpecta
admodum exigua fuit; veruntamen hora fexta
reliquis magnitudine fere fuit æqualis.

   Die 20. hora 1. min. 15. conftitutio confimilis
vifa eft. Aderant tres Stellulæ adeo exiguæ, vt vix
                                            Ori

8 *Galileo's observations of Jupiter and its four brightest
satellites, from his* Siderius Nuncius *(Florence 1610).
Supporters of the Aristotelian universe argued that if the
Earth moved, then the Moon would be left behind. By
observing the satellites orbiting round Jupiter, a planet
which all agreed to be moving through space, Galileo
believed he had observational evidence strongly supporting
the concept of a moving Earth*

the body at a uniform rate to infinity, and the velocity
which results from a natural acceleration downwards
common to all bodies.

10 Newton's 'Axioms', or 'Laws of Motion'

*What Galileo began to formulate, Issac Newton
(1642-1727) developed and completed in his famous
PHILOSOPHIAE NATURALIS PRINCIPIA MATHEMATICA
(Mathematical Principles of Natural Philosophy)
which was published in London in 1687. This was a
masterpiece of logical mathematical thinking applied to
the whole question of the movement of bodies on Earth
and in space, and to the basic concept of universal
gravitation. On the laws of motion, Newton was quite
explicit at the beginning of the book:*

LAW I

*Every body continues in its state of rest, or of uniform
motion in a right line [straight line], unless it is
compelled to change that state by forces impressed upon it.*
Projectiles continue in their motions, so far as they
are not retarded by the resistance of the air, or
impelled downwards by the force of gravity. A top,
whose parts by their cohesion are continually drawn
aside from rectilinear motions, does not cease its
rotation, otherwise than as it is retarded by the air.
The greater bodies of the planets and comets, meeting
with less resistance in freer spaces, preserve their
motions both progressive and circular for a much
longer time.

LAW II

*The change of motion is proportional to the motive
force impressed; and is made in the direction of the right
line in which that force is impressed.*
If any force generates a motion, a double force will
generate double the motion, a triple force triple the
motion, whether that force be impressed altogether
and at once, or gradually and successively. And this
motion (being always directed the same way with the
generating force), if the body moved before is added
to or subtracted from the former motion, accordingly

 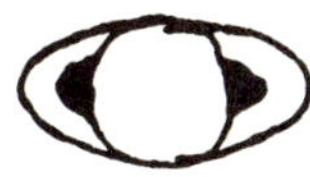 

9 *Three of Galileo's sketches of the planet Saturn, from a letter written in December 1612 to his friend Marcus Welser (1558-1614), a scholar living in Augsburg. They show what Galileo called the 'triple nature' of the planet, and which he thought might be caused by two stationary satellites. The ever changing appearance of the planet worried him because he felt that his opponents might use this to discredit his telescope and his observations*

as they directly conspire with or are directly contrary to each other; or obliquely joined, when they are oblique, so as to produce a new motion compounded from the determination of both.

LAW III

To every action there is always opposed an equal reaction: or, the mutual actions of two bodies upon each other are always equal, and directed to contrary parts.

Whatever draws or presses another is as much drawn or pressed by that other. If you press a stone with your finger, the finger is also pressed by the stone. If a horse draws a stone tied to a rope, the horse (if I may say so) will be equally drawn back towards the stone; for the distended rope, by the same endeavour to relax or unbend itself, will draw the horse as much towards the stone as it does the stone towards the horse, and will obstruct the progress of the one as much as it advances that of the other. If a body impinge upon another, and by its force change the motion of the other, that body also (because of the equality of the mutual pressure) will undergo an equal change in its own motion, towards the contrary part. . . .

11 Newton on universal gravitation

Using these laws and universal gravitation, he could then write at the end of the whole volume, after he had dealt in detail with planetary motion in elliptical orbits and terrestrial motion under all kinds of conditions:

Hitherto we have explained the phenomena of the heavens and of our sea [the tides] by the power of gravity, but have not yet assigned the cause of this power. This is certain, that it must proceed from a cause that penetrates to the very centres of the Sun and planets, without suffering the least diminution of

10 *The true nature of Saturn—that is a planet surrounded by rings—was discovered in 1655 by the Dutchman Christian Huygens (1629-95), using a telescope superior in quality to those possessed by Galileo. From Huygens'* Systema Saturnium *(Leyden 1659)*

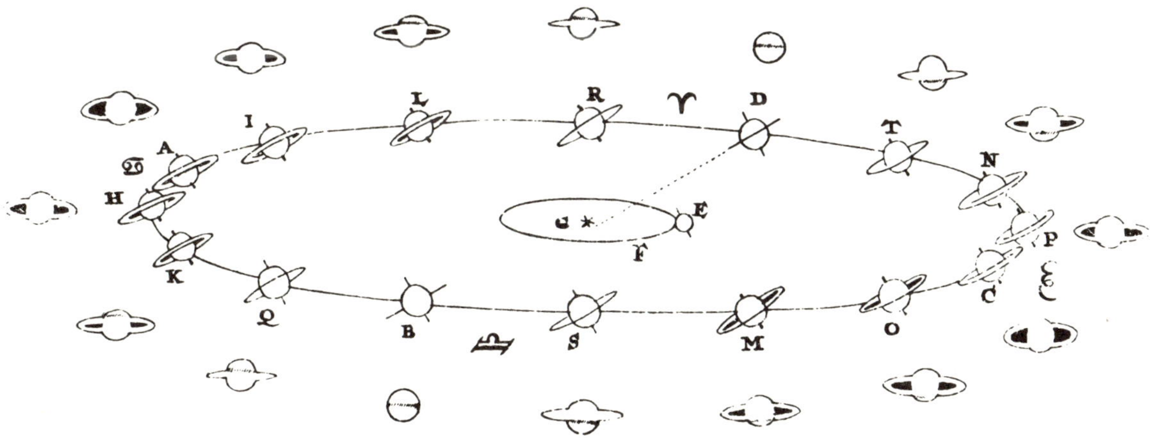

23

its force; that operates not according to the surfaces of the particles on which it acts (as mechanical causes used to do), but according to the quantity of the solid matter which they contain, and propagates its virtue on all sides to immense distances, decreasing always as the inverse square of the distances. Gravitation towards the Sun is made up out of the gravitation towards the several particles of which the body of the Sun is composed; and in receding from the Sun decreases accurately as the inverse square of the distances as far as the orbit of Saturn [then the farthest known planet], as evidently appears from the quiescence of the aphelion of the planets; nay, and even to the remotest aphelion of the comets, if those aphelions are also quiescent. But hitherto I have not been able to discover the cause of those properties of gravity from phenomena, and I frame no hypotheses; for whatever is not deduced from the phenomena is to be called an hypothesis; and hypotheses, whether metaphysical or physical, whether of occult qualities or mechanical, have no place in experimental philosophy. In this philosophy particular propositions are inferred from the phenomena, and afterwards rendered general by induction. Thus it was that the impenetrability, the mobility, and the impulsive force of bodies, and the laws of motion and of gravitation were discovered. And to us it is enough that gravity does really exist, and act according to the laws which we have explained, and abundantly serves to account for all the motions of the celestial bodies, and of our sea.

The doctrine of universal gravitation was accepted by some scientific men, but not by all, in spite of the brilliant advocacy of the Principia, *for its nature remained too much of a mystery.*

12 Halley on cometary orbits

However, Edmond Halley (1656-1742), who had persuaded Newton to write the Principia, *had edited the manuscript and paid for its publication, used the concept and the laws of motion to analyse the behaviour of comets. In the seventeenth century much time had been spent discussing cometary paths, but since comets were only visible when near the Sun, observational evidence was insufficient to enable a choice to be made between a straight line or one of the family of curves— the hyperbola, parabola, or ellipse. Newton inclined towards the parabola; but Halley decided to make a thorough investigation, publishing his results both in the* Philosophical Transactions of the Royal Society (*still the most famous of scientific periodicals*), *and in* A SYNOPSIS OF COMETARY ASTRONOMY, *which appeared in* 1705:

Hitherto I have consider'd the Orbits of Comets as exactly *Parabolick;* upon which Supposition it wou'd follow that Comets being impell'd towards the Sun by a Centripetal Force, descend as from Spaces infinitely distant, and by their Falls acquire such a Velocity, as that they may again run off into the remotest Parts of the Universe, moving upwards with such a perpetual Tendency, as never to return again to the Sun. But since they appear frequently enough, and since none of them can be found to move with an Hyperbolick Motion, or a Motion swifter than what a Comet might acquire by its Gravity to the Sun, 'tis highly probable that they rather move in very Excentric Orbits, and make their Returns after long Periods of Time: For so their Number will be determinate, and, perhaps, not so very great.

. . . And, indeed, there are many Things which make me believe that the Comet which Apian [Peter Apian (1495-1552)] observed in the year 1531, was the same with that which *Kepler* and *Longomontanus* [Christian Longomontanus (1562-1647)], took notice of and describ'd in the Year 1607, and which I my self have seen return, and observ'd in the Year 1682. All the Elements [orbital details] agree, and nothing seems to contradict this my Opinion, besides the Inequality of the Periodick Revolutions: Which Inequality is not so great neither, as that it may not be owing to Physical Causes. For the motion of *Saturn* is so disturbed by the rest of the Planets, especially *Jupiter*, that the Periodick Time of that Planet is uncertain for some whole Days together. How much more therefore will a Comet be subject to such Errors, which rises almost Four times higher than *Saturn*, and whose Velocity, tho' encreased but very little, would be sufficient to change its Orbit, from an Elliptical to a Parabolical one, . . .

Later, Halley gave further details, predicting the return of this particular comet at about the end of 1758 and, although he was by then dead, observers did indeed pick up the comet on Christmas Day of that year; it was delayed, just as Halley had foreseen, by the gravitational pull of Jupiter, although the delay he had calculated was a little short of the observed value. But this did prove his point, that comets, or at least some of them, do travel in elliptical orbits, and the periodic returns of these are predictable. And this soon disposed of the popular belief that they were evil omens.

13 William Herschel on the discovery of Uranus

Observations of planetary features were virtually impossible before the advent of the telescope, and the first

11 *Halley's comet photographed at its re-appearance in 1910 April, from Kodaikanal in India (see extract 12)*

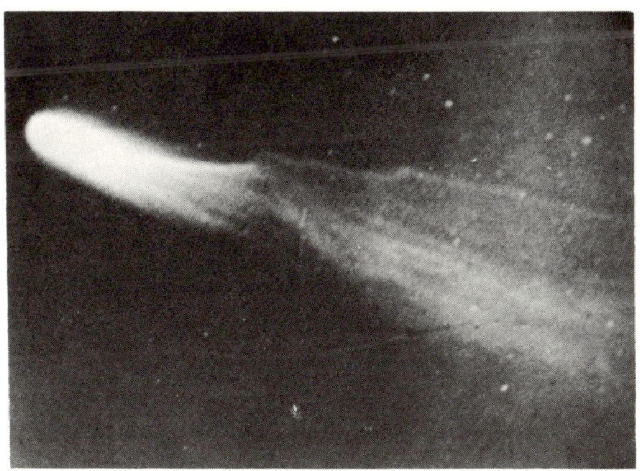

well-documented astronomical use was by Galileo. With instruments of his own design, he observed Jupiter, and discovered it possessed satellites, saw that Venus had phases similar to those of the Moon, and used both facts to drive nails into the coffin of Aristotle's and Ptolemy's ideas about the universe. But it was the development of the telescope by William Herschel (1738-1822) that in 1781 led him to the discovery of a new planet, the first to be found since the unrecorded discovery of the others in earliest times. His observations did not make everything immediately clear, for at first he thought he was observing a comet:

On Tuesday the 13th of March, between ten and eleven in the evening, while I was examining the small stars in the neighbourhood of H Geminorum, I perceived one that appeared visibly larger than the rest: being struck with its uncommon magnitude [brightness], I compared it to H Geminorum and the small star in the quartile [90°] between Auriga and Gemini, and finding it so much larger than either of them, suspected it to be a comet.

———

Miscellaneous observations and remarks

March 19. The Comet's apparent motion is at present 2½ seconds [of arc] *per* hour. It moves according to the order of the signs [same direction as the Sun], and its orbit declines but very little from the ecliptic.

March 25. The apparent motion of the Comet is accelerating, and its apparent diameter seems to be increasing.

March 28. The diameter is certainly increased, from which we may conclude that the Comet approaches to us.

———

April 6. With a magnifying power of 278 times the Comet appeared perfectly sharp upon the edges, and extremely well defined, without the least appearance of any beard or tail.

When not close to the Sun, a comet appears as a patch or disk of light, since it then has no beard or tail, and Herschel's original surmise that he was observing a comet is understandable, although his observation of April 6 introduced an element of doubt. The Astronomer Royal, Nevil Maskelyne (1732-1811), was soon informed of Herschel's observations, and he independently noted that the object had no tail but presented a sharp disk. Subsequently, calculations of its orbit showed the object to be a planet, not a comet, with an orbit lying beyond that of Saturn, and by international consent it was named Uranus.

This discovery stimulated many astronomers, and from 1801 onwards numbers of small planets were discovered by Giuseppe Piazzi (1746-1826) and others: all had orbits lying between Mars and Jupiter. But was Uranus the most outlying of all the planets?

14 Adams' intention to investigate the motion of Uranus

As the years passed, it was found that Uranus did not follow the path it should if Newton's gravitational theory were correct. Due to the work of French mathematicians such as Pierre Laplace (1749-1827), the detailed application of Newtonian celestial mechanics to the solar system seemed to leave no doubt about their veracity. Some other explanation had to be found, and two astronomers formulated an answer—the Frenchman Urbain Leverrier (1811-77) and the young English mathematical astronomer John Couch Adams (1819-92). Among Adams' papers in St John's College, Cambridge, is his note about the matter:

1841. July 3. Formed a design, in the beginning of this week, of investigating, as soon as possible after taking my degree, the irregularities of the motion of Uranus, wh. are yet unaccounted for; in order to find whether they may be attributed to the action of an undiscovered planet beyond it; and if possible thence to determine the elements of its orbit, &c. approximately wh wd. probably lead to its discovery.

15 Leverrier predicts the position of the undiscovered planet, Neptune

Leverrier, for his part, did not begin work on the problem until June 1845, two years after Adams had struggled with it. Adams obtained two solutions, the second by mid-September 1845, but neither was published. Leverrier, on the other hand, made his work known, and wrote to Heinrich Schumacher (1780-1850), editer and founder of the famous journal ASTRONOMISCHE NACHRICTEN:

No. 580
Letter from Herr Leverrier to the Editor.
Paris 8 September 1846.

In the last letter that I had the pleasure of writing to you, I reported to you that I had undertaken some extensive researches on the movements of Uranus; and that I was reaching the conclusion that a perturbing planet existed, of which I indicated the position. I have been very busy since then, perfecting my calculations, and I formed the desire to reach a conclusion before the time of opposition [on the opposite side of the Earth to the Sun] of the new star, so that astronomical observers could explore with ease the region of the sky brought to their attention...

I take the liberty of sending you an extract from my work, with the request that you insert it in your learned journal. I hope before long to be able to publish my researches in detail. May they inspire enough confidence in astronomical observers to encourage them to study carefully the part of the sky where it will without doubt be possible to discover the planet, the mass of which is considerable.

16 John Herschel on the discovery of Neptune

The delay in the announcement of Adams' results was due to a series of unfortunate procrastinations by the Astronomer Royal, George Biddell Airy (1801-92) and James Challis (1803-82) of the Cambridge Observatories, and in the end Leverrier's solution was the first to be confirmed by observation. And it is to

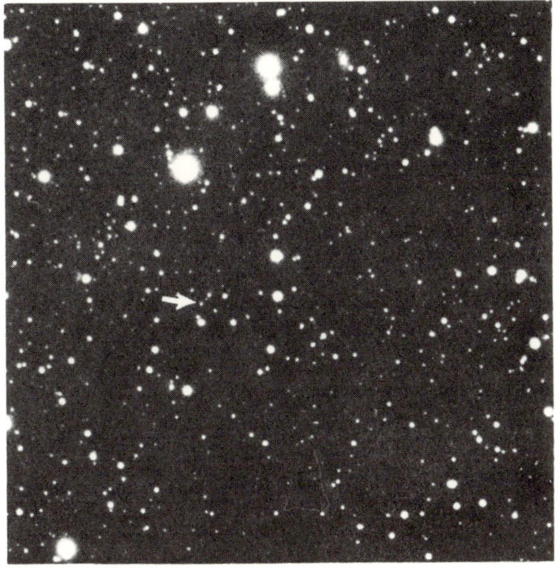

Leverrier alone that the discovery would have been credited, had not William Herschel's son John Herschel (1792-1871), used his immense international standing and written to the ATHENAEUM *magazine at the beginning of October* 1846:

. . . The remarkable calculations of M. Le Verrier which have pointed out, as now appears, nearly the true situation of the new planet by resolving the inverse problem of the perturbations [of Uranus]—if uncorroborated by the numerical calculations of another hand, or by independent investigation from another quarter—would hardly justify so strong an assurance as that conveyed by my expressions above alluded to [his address as retiring President of the British Association for the Advancement of Science at Southampton a month before]. But it was known to me at that time (I will take the liberty to cite the Astronomer Royal as my authority) that a similar investigation had been independently entered into, and a conclusion as to the situation of the new planet very nearly coincident with M. Le Verrier's arrived

12 *Two photographs, one taken on 1930 January 23 and the other on January 29 at the Lowell Observatory, Flagstaff, Arizona, by Clyde Tombaugh. Pluto, which appears just like an insignificant star not a planet, is arrowed; its movement across the background of stars led to its detection*

at (in entire ignorance of his conclusions) by a young Cambridge mathematician, Mr. Adams, who will, I hope, pardon this mention of his name (the matter being one of great historical moment), and who will doubtless in his own good time and manner, place his calculations before the public.

A great controversy arose over the validity of Adams' claim, but in the end it became appreciated that both he and Leverrier had independently solved the very difficult problem they had so boldly tackled. As time went on, however, it was found that Neptune itself was perturbed, and that its presence did not account for all the perturbations suffered by Uranus. A hunt therefore began for a planet still further from the Sun, and some possible positions were computed by Percival Lowell (1855-1916) and, appropriately enough, the planet was discovered in

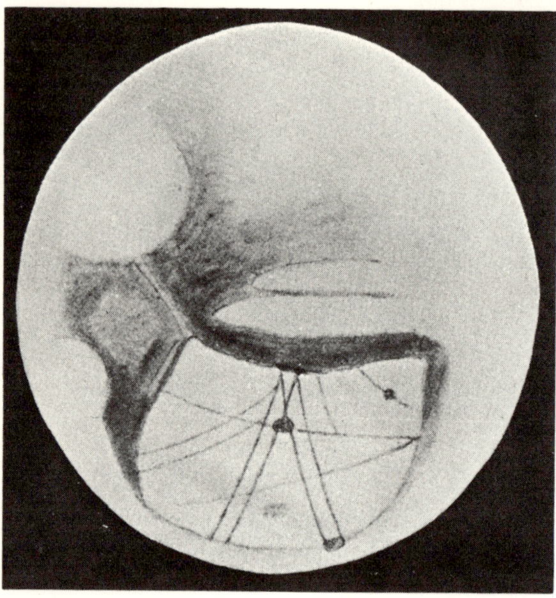

13 *A drawing by Percival Lowell (1855-1916) of the surface of Mars, showing the 'canals' crossing all over the planet and which Lowell believed he had clearly observed. From his book* Mars (*London 1896*)

1930 at the observatory Lowell founded, by Clyde Tombaugh (1906-   ). Yet the planet appears to be smaller than would be required for it to account for the observed perturbations and, to some astronomers, the outermost planet of the solar system, Pluto, and the perturbations of Uranus and Neptune are still not satisfactorily explained.

A new era in planetary research opened in October 1957 when the Russians successfully launched the first artificial satellite into orbit around the Earth. Now, a little more than a decade later, it is possible to probe the Moon and planets in ways hitherto the province of science fiction, ways that are of great significance for astronomy. With the landing of Man on the Moon in July 1969, the manned exploration of the solar system has begun, and again the results hold immense promise, as the following photographic evidence shows.

14 *The surface of Mars which, in reality, possesses no canals, photographed by the American space-probe Mariner 6 in 1969 July when only 2,150 miles from the planet. The area covered in this photograph is more than the area of France: the large crater on the right has a diameter of about 170 miles, and the whole terrain is reminiscent of the surface of the Moon*

28

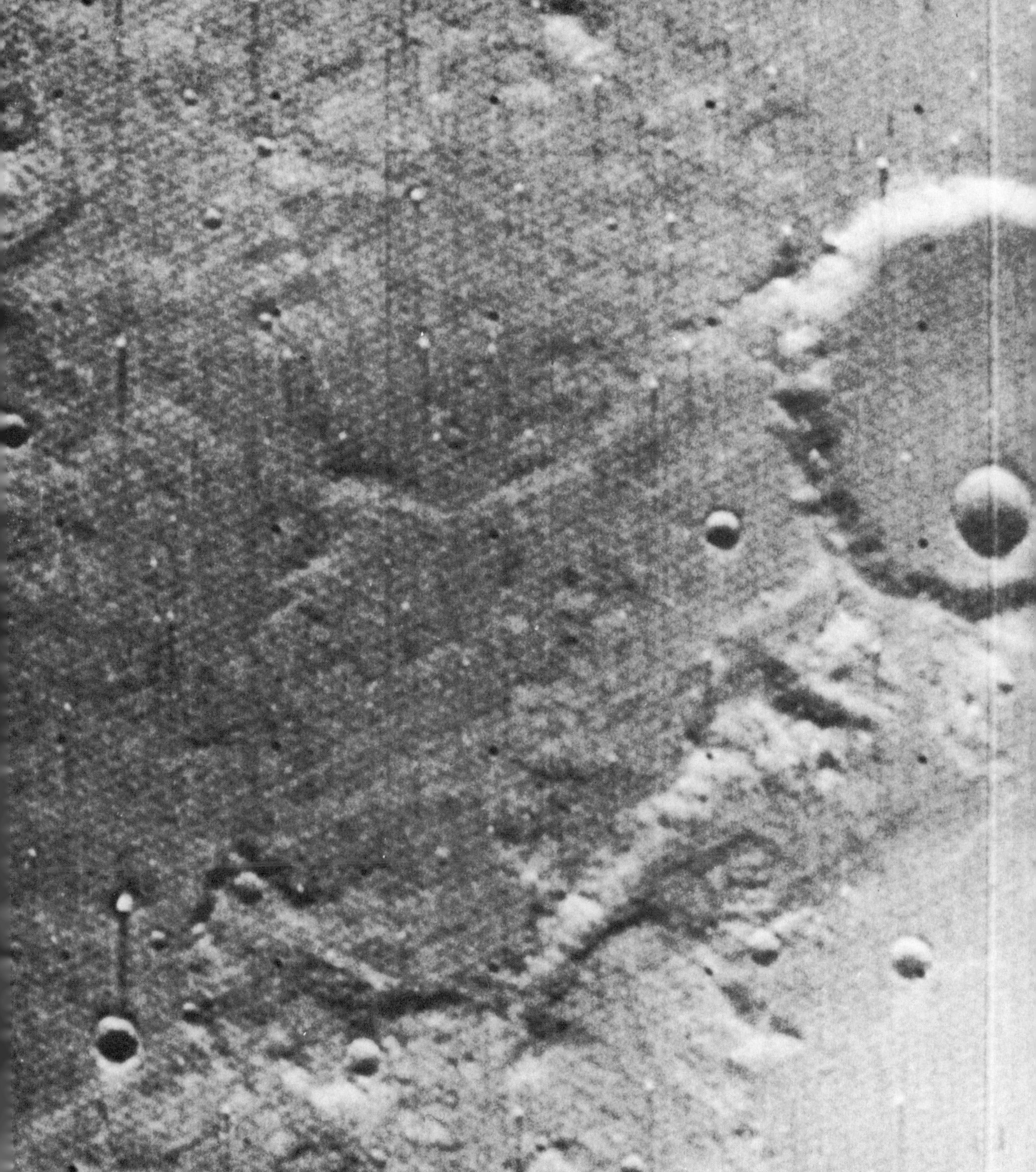

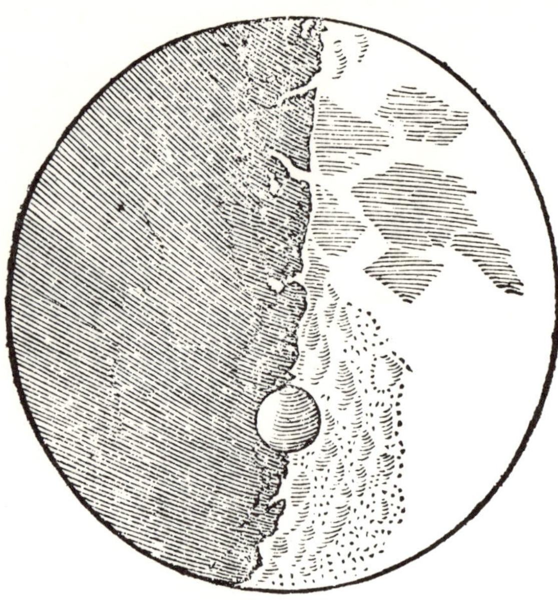

15 *Galileo's drawing of the Moon, showing surface features and, hence, its similarity to the Earth. This appeared in his* Siderius Nuncius, *and was the first telescopic picture of the Moon to be published*

16 *A map of the lunar surface prepared by Johannes Hevelius (1611-81), a brewer of Danzig who had an international reputation as an astronomer. From his* Selenographia *published at Danzig in 1647. The improvement in detail made possible by advances in telescope optics since the Siderius Nuncius will be obvious*

REGIO HYPERBOREA
MARE HŸPERBOREVM
MARE MEDITER
MARE ADRIATICVM
Mare Pampulium
ROMANIA
PROPONTIS
PONTVS
EVXINVS
PALVS MOEOTIS
CHERSONESVS
COLCHIS
CASPIVM
ASIATICA P.
SARMATIA
PERSIA
EXPLICATIO LITERARVM.
Mons . I. Insula
Sinus . L. Lacus
Palus . C. Caput
Promontorium .
Paludes Fl. Fluvius
vel F. Frelūm
Digiti Ecliptici et eorum Segmenta.

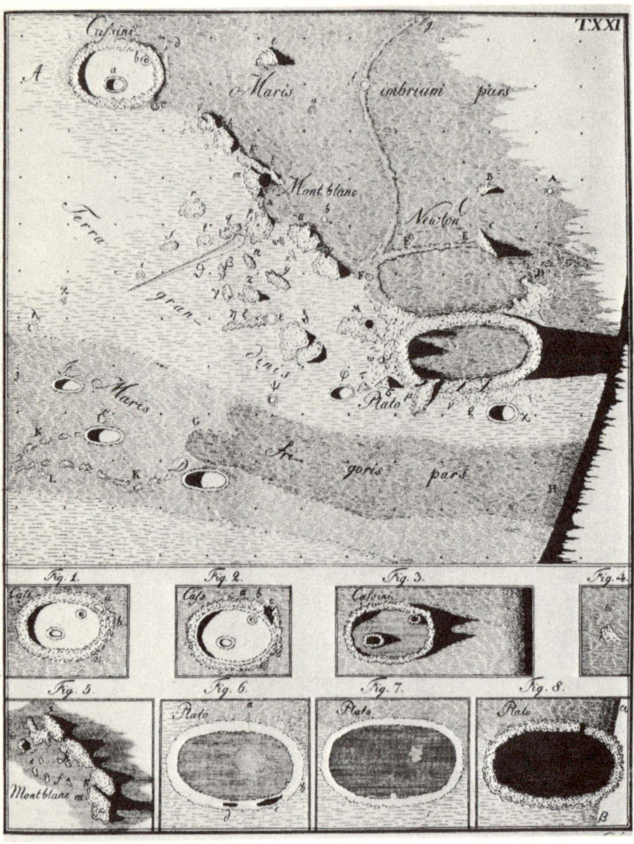

19 *Photograph of the lunar surface around the Mare Imbrium taken with the 200in aperture reflector at Palomar Observatory in California. This is the most detailed picture that can at present be obtained from an Earth-based observatory (right)*

17 *Details of the lunar surface near the Mare Imbrium drawn by Johann Schröter (1745-1816) at his private observatory near Bremen. From his* Selenographische Fragmente *published in 1791 at Göttingen*

18 *Rear of the Moon photographed by the Russian space-probe Lunik 3 in 1959. Although better pictures have since been obtained, this photograph was Man's first view of the face of the Moon that remains perpetually turned away from the Earth. It shows that the far side possesses fewer maria ('seas') than the side that faces us*

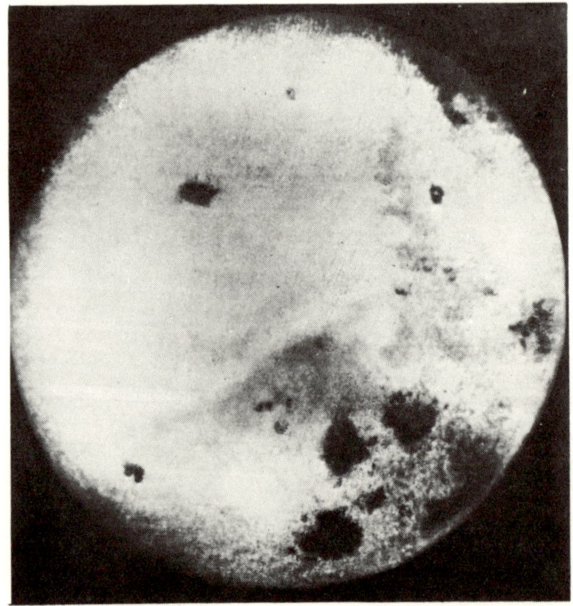

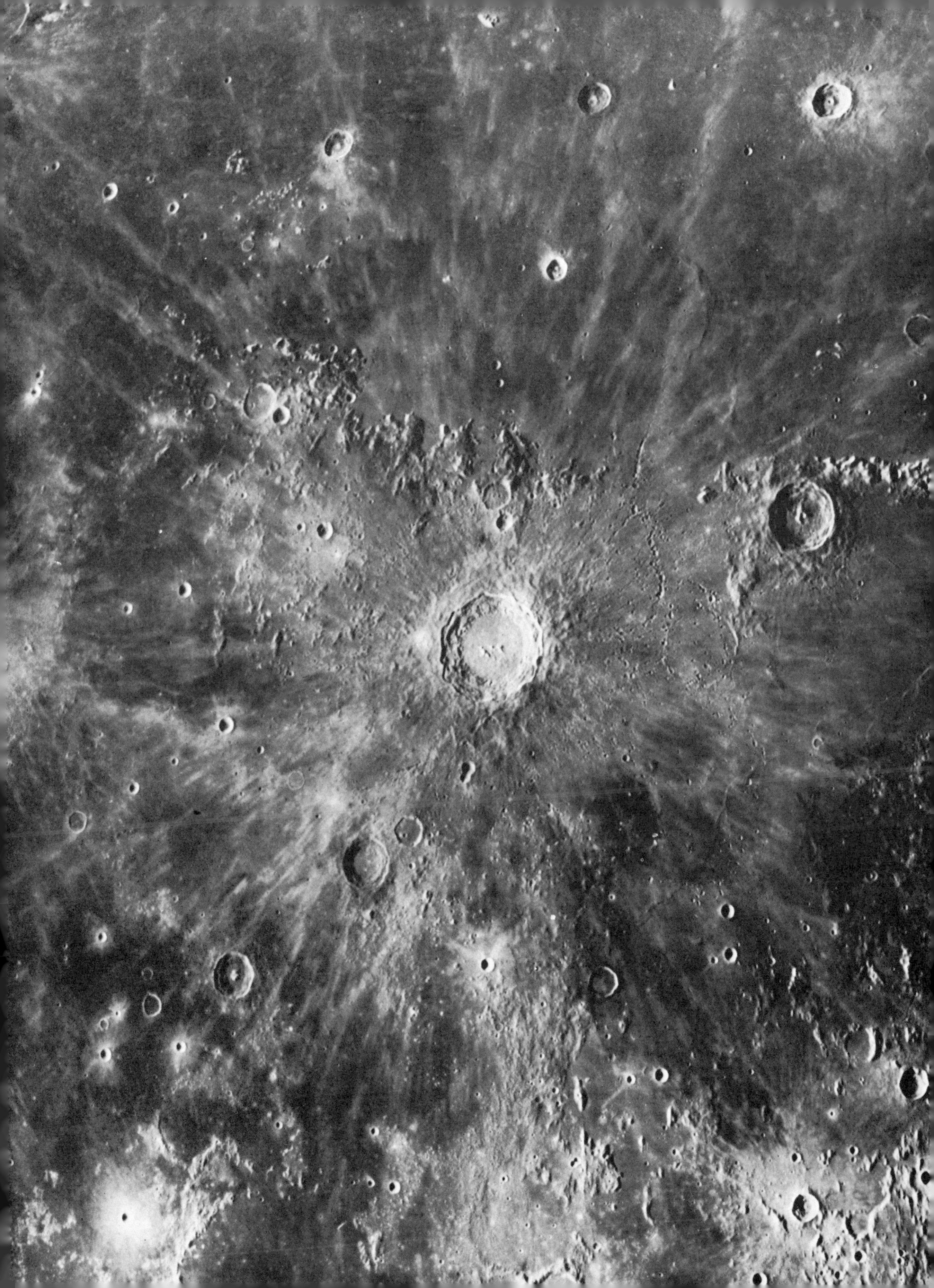

20 *Now the surface of the Moon can be mapped by orbiting space craft and explored by remote-controlled machines and by Man. This photograph, taken in July 1969, not only shows a close-up of some of the lunar surface but also the first astronauts to land on another body in space. Here Edwin Aldrin has just set up a seismometer to measure 'moon-quakes' on the Mare Tranquillitatis, as well as a mirror to reflect a laser beam radiated from the Earth back to observers there. The lunar excursion module of* Apollo 11 *is seen in the background (right)*

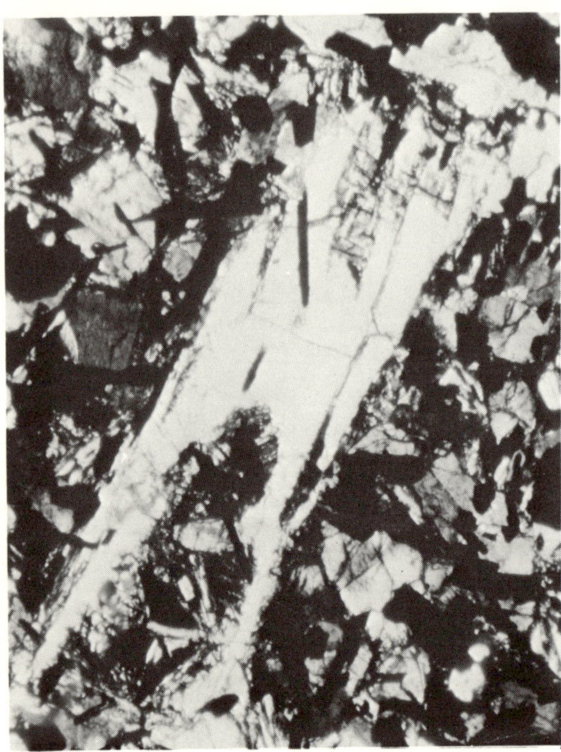

21 *A photomicrograph, taken in polarised light, of a thin section of a sample of lunar rock brought back from the Mare Tranquillitatis by the crew of* Apollo 11, *showing the minerals plagioclase, ilmenite and pyroxene*

CHAPTER TWO

Measuring the Stars

Astronomy is the oldest of the exact sciences. Precision is an integral part of it, and nowhere does this show more clearly than in the measurement of star positions. Here meticulous care is needed, and throughout the ages astronomers with a flair for precise observation have stretched contemporary techniques to their limit, making measurements that have notably advanced the whole science, sometimes in unexpected ways.

In the earliest days of astronomy, the stars were thought of as studded on the inside of the giant sphere of the sky, and although we now know this is untrue, it is still a convenient fiction to adopt when it comes to specifying the relative position of one star to another in the sky. This is because all celestial measurements have to be made as angles, since it is the only convenient and satisfactory way to specify relative positions, and the actual units used are still those devised by the Babylonians some two millenia ago. The whole circle of the sky is divided into 360 degrees, and each degree (°) sub-divided into 60 minutes (') of arc, and each minute

into 60 seconds (") of arc.

Having decided on the units, the astronomer must next specify in which directions the measurements are to be made, and two possibilities present themselves. In the first place, if one thinks of the Earth as fixed in the centre of the sphere of the stars, then one can imagine the equator and the poles projected outwards on to the sphere (figure 22). We then have the 'celestial equator' and the celestial poles, the whole sphere of heaven appearing to rotate once a day about these poles. Units measured eastwards along the celestial equator are known as units of 'right ascension' and not celestial longitude as might be imagined from an analogy with terrestrial longitude: north or south to the celestial poles is not celestial latitude either, but is called 'declination'.

The terms celestial longitude and celestial latitude are used for a different set of co-ordinates, originally adopted when astronomy and her sister, the pseudo-science of astrology, were closely linked. These are measured eastwards along the ecliptic, and north and

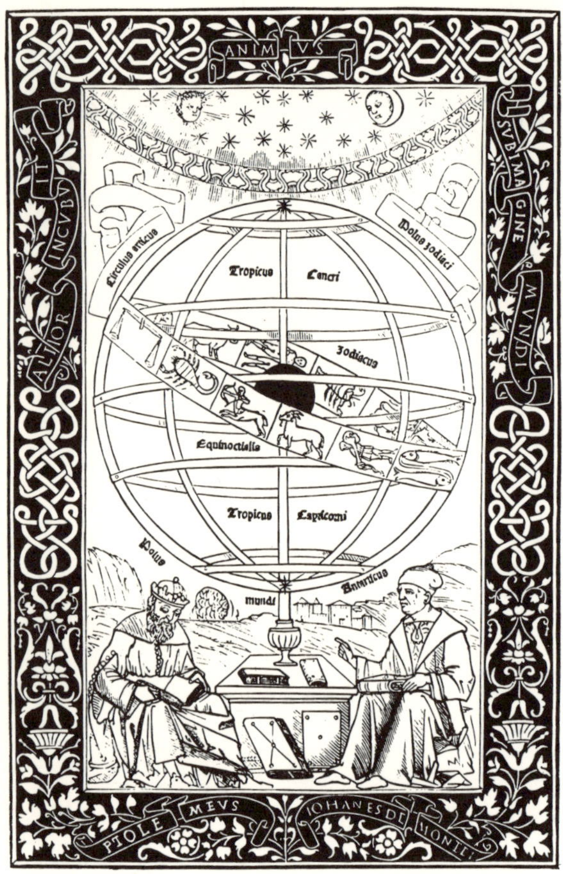

22 *The frontispiece of the* Epitome in Ptolemæi Almagest *(Epitome of Ptolemy's Almagest) by Regiomontanus, published in Rome in 1496. The author of this highly successful treatise on Ptolemaic astronomy was Johann Müller (1436-76), who was born in Königsberg from which he took the name Regiomontanus*

south of it (figure 22). The ecliptic is close to the Moon's path and those of all the planets, and passes through all the signs of the zodiac, so beloved of astrologers.

17 Ptolemy's star catalogue

The starting point for measuring celestial longitude or *for right ascension is one of the two points where the ecliptic cuts the celestial equator, the one chosen being that where the Sun moves from south to north of the equator because this coincides with the coming of spring in the northern hemisphere—it is the 'vernal equinox'. These two intersecting points are not fixed in space and each year move backwards (westwards) by a very small amount, a fact that was discovered by the Greek astronomer Hipparchos about 150 BC, when he compared star positions measured many centuries apart. Known as the 'precession of the equinoxes', it was familiar to Ptolemy, who allowed for it in the star positions in his catalogue in* ALMAGEST:

TABULAR EXPOSITION OF THE CONSTELLATIONS OF THE NORTHERN HEMISPHERE

Configurations	*Longitude*	*Latitude*	*Magn.*
Constellation of Ursa Minor (Smaller Bear)			
The star on the tip of the tail	Twins $0°\frac{1}{6}'$	N66°	3
The next one in the tail ..	Twins $2\frac{1}{2}°$	N70°	4
The next one, before the beginning of the tail ..	Twins 16°	N$74\frac{1}{3}°$	4
The southern one on the western side of the rectangle	Twins $29\frac{2}{3}°$	N$75\frac{2}{3}°$	4
The northern one on the same side	Crab $3\frac{2}{3}°$	N$77\frac{2}{3}°$	4
The southern one of those on the eastern side ..	Crab $17\frac{1}{2}°$	N$72\frac{5}{60}°$	2
The northern one on the same side..	Crab $26\frac{1}{6}°$	N$74\frac{5}{60}°$	2
In all, 7 stars of which 2 are of 2nd magnitude, 1 of 3rd, 4 of 4th.			
The unfigured star near it, that is the more southern star of first magnitude in a straight line with those on the eastern side	Crab 13°	N$71\frac{1}{6}°$	4

Septima 78

¶ Longitudo et Latitudo ac Magnitudo stellarum fixarum

¶ Forme et Stelle	ǒ	g	m		g	m	mag
¶ Stellatio Urse Minoris Imago Prima	Lōgitudo				Lat.º		
Illa que est super extremitatem caude	2	0	10	S	66	0	3
Illa que est post istam super caudam	2	2	30	S	70	0	4
Illa que est post eam in origine caude	2	16	0	S	74	0	4
Meridiana a latere antecedente laterum clunium	2	29	40	S	75	40	4
Septentrionalis ab hoc latere	3	3	40	S	77	40	4
Meridiana duarum que sunt in latere sequente	3	17	10	S	72	50	2
Septentrionalis ab hoc loco	3	26	10	S	74	50	2
De ergo sunt septem stelle, quarum in magnitudine secuda sunt due in tertia vna. z in qrta qttuor.							
¶ Que est inter eas: z non est in forma.							
Meridiana duax q st sup rectitudine duax stellax q sut i late seqnte	3	13	0	S	71	10	4
¶ Stellatio Urse Maioris Imago Secunda							
Illa que est super extremitatem muscide.	2	25	20	S	39	50	4
Antecedens duarum que sunt in duobus oculis.	2	25	50	S	43	0	5
Sequens earum	2	26	20	S	43	0	5
Antecedens duarum que sunt in fronte	2	26	10	S	47	10	5
Sequens earum	2	27	40	S	47	0	5
Illa que est super extremitatem auris antecedentis	2	28	10	S	50	30	5
Antecedens duarum que sunt in collo	3	2	30	S	43	50	4
Sequens earum	3	9	30	S	44	20	4
Declinior duarum earum que sunt in pectore ad septentrionem	2	11	0	S	42	0	4
Declinior earum ad meridiem	2	10	0	S	44	0	4 e.t.
Illa que est super genu sinistrum	3	5	40	S	35	0	3
Septetrionalis duax stellaru q sut i extremitate pedis sinistri pcedet	3	6	30	S	29	20	3
Meridiana earum	3	5	20	S	28	30	3
Illa que est super genu dextrum	3	5	40	S	36	0	4
Illa que est sub genu dextro	3	17	50	S	33	3	4
Illa que est super dorsum earum que sunt habentis quattuor latera	3	22	40	S	49	0	2
Illa que est super mirach eius	3	2	10	S	44	30	2
Illa que est super originem caude eius	4	3	10	S	51	0	3
Sequens earum: z est illa q est super ancham sinistram posteriorem	4	4	0	S	46	30	2
Antecedens duarum que sunt in pede sinistro posteriore	3	22	40	S	29	30	3
Sequens hanc	3	24	10	S	28	15	3
Illa que est in ventre genu sinistri	4	1	40	S	35	15	4
Septentrionalis duaru que sunt in pede dextro posteriore	4	9	50	S	25	50	3
Declinior earum ad meridiem	4	10	20	S	25	0	3
Prima trium que sunt super caudam: z est aliore	4	12	10	S	53	30	2
Media earum	4	18	0	S	55	40	2
Tertia: z est ea que est super extremitatem caude	4	29	50	S	54	0	2
Illarum ergo vigintiseptem stellarum in magnitudine secunda sunt sex. in tertia octo. in quarta octo. in quinta quinq.							
¶ Ille que sunt sub eis z non sunt in forma							
Stella elongata a cauda versus meridiem	4	27	50	S	39	45	3
Antecedens hanc: z est occultior ea	4	20	10	S	41	20	5
Declinior duarum que sunt in eo qd est inter duos pedes antecedentes vrse z inter caput leonis ad meridiem.	3	15	0	S	17	35	4
Illa que est declinior ab hac ad septentrionem	3	13	20	S	19	10	4
Sequens stellarum trium reliquarum occultarum	3	16	10	S	20	0	oc.
Antecedens hanc	3	12	10	S	22	45	oc.
Illa que plus antecedit hanc	3	11	10	S	23	15	oc.
Illa que est inter duos pedes vrse antecedentes z geminos	4	0	0	S	22	15	oc.
Illarum ergo octo stellarum que non sunt in forma: in magnitudine tertia est vna. in quarta due. in quinta vna. z occulte quattuor.							
¶ Stellatio Draconis Imago Tertia							
Que est super linguam	6	26	40	S	76	30	4
Que est in ore	7	11	50	S	78	30	4

23 *A section of Ptolemy's catalogue of stars from his* Almagest, *photographed from the edition of 1515 published in Venice. An English translation is given in extract 17*

Ptolemy gives his celestial longitudes from the beginning of each zodiacal sign which, by convention, occupies a longitude of 30°. The whole catalogue contains 1,022 stars and, as will be seen, the smallest angle measured is

$\frac{1}{6}$ *of a degree or ten minutes of arc* (10′), *although his real accuracy of position was, by present standards, only half as good as this. The last column gives the apparent brightness or 'magnitude' (because a brighter star looks bigger to the eye): on his magnitude scale the brightest stars were denoted by* 1 *and the dimmest by* 6.

18 Tycho Brahe's star catalogue

As the years went by, and the torch of Greek learning was kept alight by the Muslim world, Ptolemy's catalogue was copied and re-copied, corrections for precession being made each time the star list was drawn up. Not until the fifteenth century were star places

24 *An early observing instrument that became known to western astronomers as 'Ptolemy's rules', although its invention seems certainly to have pre-dated Ptolemy. Hipparchos probably used it, and perhaps also the plinth (left) which, set up in a north-south position, showed by the shadows cast, when the Sun lay due south. In using the rules, the stand was placed level and sightings made along the arm GD. The position to which this arm had to be moved was a measure of the altitude of a celestial body above the horizon, the degree of altitude being shown on the scale along the arm FE. From Tycho Brahe,* Astronomiae instaurate mechanica (The Mechanics of the Renewed Astronomy) *(Prague 1602)*

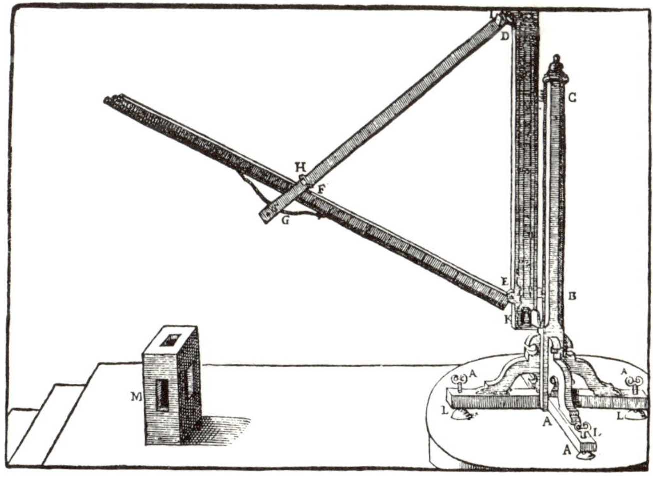

freshly observed and catalogued, when Ulugh Beg, a grandson of Tamerlane, established an observatory at Samarkand with huge instruments made of stone. Unfortunately, Ulugh Beg was the victim of a political murder and his catalogue did not become known until it was printed in 1665, but by then it had already been superseded in accuracy by new catalogues produced in western Europe. Of these Tycho Brahe's was the most exact, with a degree of precision twenty times as good as Ptolemy's, for compared with modern values, his star positions were correct to within 4' (although he catalogues down to $\frac{1}{6}$'). It was this improved accuracy that allowed Kepler to obtain his results on elliptical orbits:

URSA MINOR (SMALLER BEAR)

		Distance between the stars	Meridian altitude	Right ascension Declination	Longitude (observed) Latitude	Longitude (tabulated) Latitude	Magnitude
Pole star	Tail of the Swan	44 40$\frac{1}{3}$	51 15 above	5 46	22 48$\frac{3}{8}$ ♊	21 25 ♊	
	To the hip of Cassiopaeia	28 35$\frac{3}{4}$					2
	Capella	43 23$\frac{1}{2}$	48 23 below	87 4 N	66 1$\frac{1}{4}$ N	66 0 N	
The following one in the tail	Capella	47 21			25 30 ♊	23 45 ♊	4
	Schedar	34 35$\frac{2}{3}$			69 51$\frac{2}{3}$ N	70 9 N	
In the next place in the tail	Eye of the Dragon	29 54			3 22 ♋	7 15 ♋	4
	To the hip of Cassiopaeia	39 11$\frac{1}{3}$			73 49$\frac{1}{3}$ N	74 20 N	
Northern following point (of body)	Tail of the Swan	46 15$\frac{1}{2}$	65 29 above	223 17$\frac{2}{3}$ ♌	7 23$\frac{1}{2}$ ♌	8 25 ♌	3
	To the hip of Cassiopaeia	43 48	37 9 below	75 50 N	72 56$\frac{2}{3}$ N	72 50 N	
Southern preceding point (of body)	Tail of the Swan	44 32$\frac{3}{4}$	68 0 above	230 35$\frac{1}{2}$	15 32$\frac{2}{3}$ ♌	17 25 ♌	2
	To the hip of Cassiopaeia	45 7$\frac{1}{3}$	34 38 below	73 19 N	75 12$\frac{2}{5}$ N	74 50 N	
Northern preceding point (of body)	Tail of the Swan	27 14$\frac{2}{3}$	62 18 above	240 11$\frac{5}{6}$	21 30 ♋	20 55 ♋	4
	To the hip of Cassiopaeia	39 49	40 20 below	79 1 N	75 2$\frac{1}{2}$ N	75 40 N	

		Distance between the stars	Meridian altitude	Right ascension Declination	Longitude (observed) Latitude	Longitude (tabulated) Latitude	Magni-tude
Southern following point (of body)	Eye of the Dragon	24 30			24 28 ♋	24 55 ♋	
	To the hip of Cassiopaeia	39 49			77 45 N	77 40 N	4
In a position in a straight line with rear point					4 15 ♌		
					71 10 N		4

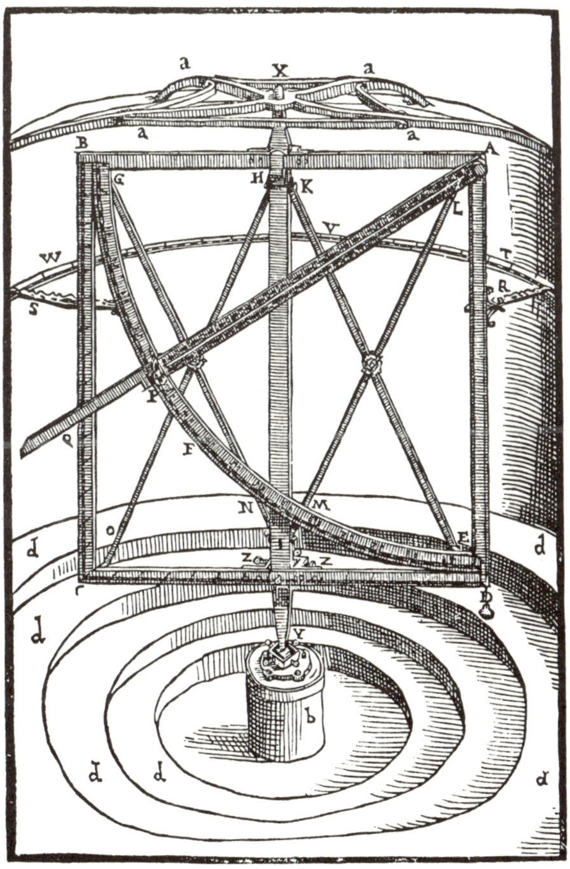

25 *A steel quadrant used by Tycho Brahe. Constructed in his workshops on the Island of Hveen (Hven) given him by Frederick II of Denmark, this, like many of his instruments, was sunk in a pit in the ground to help offset the errors introduced by strong winds. From the same source as illustration 24*

Tycho's positions were so accurate partly because of the way his instruments were constructed and mounted (figures 25 and 26), and partly because he was the first to appreciate that errors must still exist even in the most finely built instruments, and to devise an observing technique to make allowance for this. In fact, Tycho reached close to the limit of what is possible with the eye alone, for once the telescope had been invented, a device was available that could resolve detail beyond the reach of the unaided eye, as anyone who has used even a pair of binoculars knows.

19 John Flamsteed's star catalogue

Between 1676 and 1719 John Flamsteed (1646-1719) made telescopic observations that allowed him to compile a catalogue with positions accurate to at least 10″, an improvement of more than 24 times that achieved by Tycho:

In the List of Ptolemy	Tycho	Designation of STAR	Bayer chart	Right Ascension ° ' "	North Polar Distance ° ' "	Longitude S ° ' "	Latitude ° ' "	Variation Right Ascension ' "	Variation Declination to Pole ' "	Magnitude
I	I	At the end of the Tail. Polaris	α	8 17 30	2 21 20	♊24 14 43	66 4 10N	161 28	23 35	3
				9 39 30	3 26 30	21 43 37	65 42 54	133 33	23 30	6
		Another		211 24 0	13 56 55	♌ 7 9 4	70 4 54N	4 33	20 24	6
			b	213 46 30	11 0 40	♋28 23 24	70 29 38	14–30	19 49	5
Inf. 1			a	217 15 0	12 56 20	♌ 4 1 9	71 25 20	8–46	19 0	4
5	6	North of the preceding side □i	β	221 10 30	16 45 30	16 17 33	72 31 44	2 20	18 0	7
				223 5 30	14 35 20	8 55 8	72 58 26	8–27	17 25	3
				224 47 0	13 2 20	3 26 50	73 6 34	18–39	16 55	6
				225 9 0	17 2 15	17 19 48	73 40 40	2–10	16 52	7
				226 57 0	14 56 10	9 39 20	73 59 44	10–53	16 17	7
				229 26 0	17 3 50	17 14 17	74 56 0	4–40	15 30	5
					17 40 20					7
4	7	South of the preceding side □i	γ	230 24 0	17 4 20	17 10 28	75 13 1	5–26	15 13	3
					15 27 20					
7	4	North of the following side □i	θ ξ	235 17 30	11 49 50	♋26 22 26	74 45 48	39–53	13 29	5
				239 4 30	11 17 40	23 2 25	75 6 50	49– 0	12 6	4
				240 51 30	13 33 55	29 47 50	76 42 40	32– 8	11 35	7
					9 3 0					6
				245 8 0	13 22 30	26 29 28	77 24 15	36–51	9 57	5
				245 15 30	14 3 0	28 51 45	77 50 55	32–14	9 54	6
6	5	South of the following side □i	η	246 20 0	13 33 40	26 18 22	77 44 15	36–21	9 28	5
3	3	The next place in the tail	ε	259 33 0	7 32 45	4 55 25	73 53 8	120–42	4+58	4
2	2	In the tail close to the first (α)	δ	287 43 0	3 35 10	♊26 49 30	69 54 10	318–45	6—20	3
				288 9 0	3 14 30	27 6 23	69 34 25N	359–10	6—15	6.7

N.B. Those variations in Right Ascension with the sign —are negative; ie, with the given Right Ascension taken away, the remainder is positive and should be added.

Flamsteed adopted a new and more convenient way of referring to stars which had been devised in 1603 by the lawyer Johann Bayer (1572-1625), who had drawn up some charts of the stars based on Tycho's catalogue. Bayer had designated the brighter stars in each constellation by Greek letters, beginning with alpha (α) and, when he had run out of these, had used Roman letters. This is given under the column headed 'Bayer Chart' above.

20 Edmond Halley on the real motions of the stars

So far, all cataloguers had assumed that the stars were truly fixed in space—only the Sun, Moon, planets and comets moved. And there was no reason to suspect otherwise. But in 1718, Edmond Halley began to compare old star catalogues with more recent observations with a view to obtaining a better value for precession, and he chanced upon an astounding discovery:

Having of late had occasion to examine the quantity of the Precession of the Equinoctial Points, I took the pains to compare the Declinations of the fixt Stars delivered by *Ptolemy*, in the 3*d* Chapter of the 7*th* Book of his *Almag.* as observed by *Timocharis* and *Aristyllus* near 300 Years before *Christ*, and by *Hipparchus* about 170 Years after them, that is about 130 Years before *Christ*, with what we now find: and by the result of very many Calculations, I concluded that the fixt Stars in 1800 Years were advanced somewhat more than 25 degrees in Longitude, or that the Precession is somewhat more than 50″ *per ann.* . . . But while I was upon this Enquiry, I was surprized to find the Latitudes of three of the Principal Stars in Heaven directly to contradict the supposed greater *Obliquity* of the *Ecliptick*, which seems confirmed by the Latitudes of most of the rest: they being set down in the old Catalogue, as if the Plain of the Earths Orb had changed its Situation, among the fixt Stars, about 20′ since the time of *Hipparchus*. . . .

Nor are these errors of Transcription, but are proved

26 *Tycho Brahe, his assistants, and his large mural quadrant. Built on a specially constructed wall to lie in a north-south direction, the instrument was designed to give altitudes of the Sun and other celestial bodies when due south (ie on the meridian). The illustration, from the same source as figure 24, shows the method Tycho used to divide his scales to permit of greater accuracy in reading positions; this was the method of transversal points. In the background other instruments are illustrated: these include his large celestial globe and, below, an alchemical laboratory*

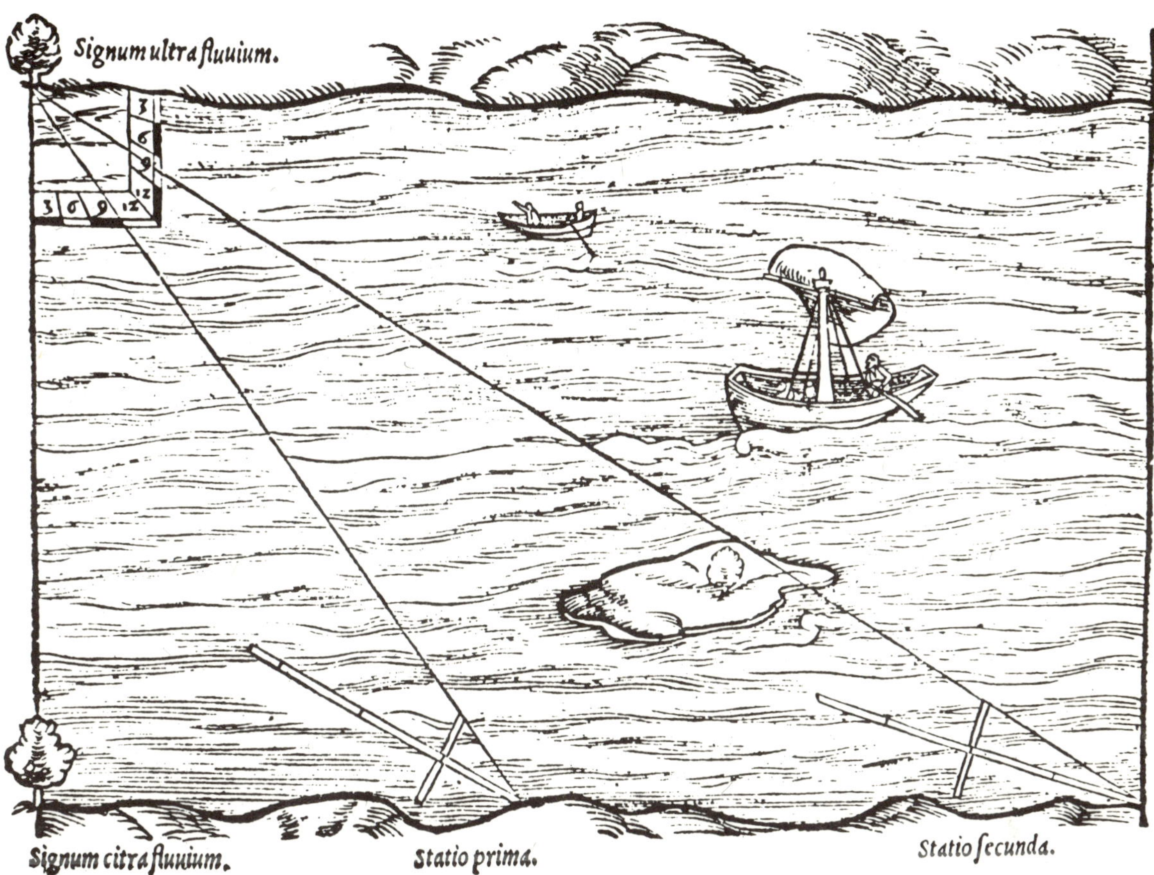

to be right by the declinations of them set down by *Ptolomy,* as observed by *Tomicharis, Hipparchus* and

27 *Flamsteed's sextant with telescopic sights instead of the open sights of earlier times. The sextant was mounted 'equatorially' so that the curved path of the stars across the sky could be followed with one motion only, instead of having to carry out two (one east-to-west—in azimuth— the other upwards or downwards in altitude) as the usual 'alt-azimuth' mounting required. Built by Thomas Tompion (1639-1713), the well-known clockmaker, it was of iron on a wooden framework, and first used in 1676. From John Flamsteed,* Historia Coelestis Britannica *(London 1725)*

28 *The method of triangulation, shown here with the angles measured using cross-staffs, in which the angle was found from the position of the cross-piece on the longer arm, the observer sighting from the end of the longer arm and over the edge of the cross-piece. From Sebastian Munster,* Rudimenta Mathematica *(Basle 1551)*

himself, which so that those Latitudes are the same as those Authors intended. . . . So then all these three Stars are found to be above half a degree more *Southerly* at this time than the Antients reckoned them. When on the contrary at the same time the bright Shoulder of *Orion* has in *Ptolomy* almost a degree more *Southerly* Latitude than at present. What shall we say then? It is scarce credible that the Antients could be deceived in so plain a matter, three Observers confirming each other. Again these Stars being the most conspicuous in Heaven, are in all probability the nearest to the earth, and if they have any particular Motion of their own, it is most likely to be perceived in them, which in so long a time as 1800 Years may shew it self by the alteration of their places, though it be utterly imperceptible in the space of a single Century of Years.

These changes were small but real enough, yet the actual velocities of the stars were unknown, and would remain so until the distances of the stars were determined. In principle this was easy, the method being based on the technique of triangulation already used by surveyors and devised in 1533 by Gemma Frisius (1508-55) (figure 28). It is essential that the places from which observations are made shall be as far apart as practicable, so that the angle to the distant object—the parallax—may be as large as possible.

21 James Bradley on the aberration of starlight

In the seventeenth and early eighteenth centuries, using the Earth's movement round the Sun to give a big displacement between observations, measurements were attempted by Robert Hooke (1635-1703) in London, and Jacques Cassini (1677-1756) in Paris, but the values they obtained seemed too large. That Cassini's results were not, in fact, measures of parallax was demonstrated by Halley, but his protégé James Bradley (1692-1762) decided to look again at Hooke's observations of the star γ (gamma) Draconis, with the help of a well-to-do friend Samuel Molyneux (1689-1728). This star appears almost overhead at the latitude of London, and they used a specially constructed long vertical telescope that should be less likely to have errors due to any bending of its tube—a common failure of early astronomical telescopes, which were very long because the light could only be brought gradually to a focus. Bradley and Molyneux obtained some interesting results, but they turned out to be due not to parallax but to a new and unexpected phenomenon—aberration:

... I have before taken notice, that it appeared from Mr. *Molyneaux's* Observations, that γ [gamma] *Draconis* altered its Declination about twice as much as the fore-mentioned small Star almost opposite to it [a star of similar general declination but on the far side of the sky]; ... This made me suspect that there might be the like Proportion between the *Maxima* of other Stars; but finding that the Observations of some of them would not perfectly correspond with such an Hypothesis, and not know, whether the small Difference I met with, might not be owing to the Uncertainty and Error of the Observations, I deferred the farther Examination into the Truth of this Hypothesis, till I should be furnished with a Series of Observations made in all Parts of the Year; ...

When the Year was compleated, I began to examine and compare my Observations, and having pretty well satisfied my self as to the general Laws of the *Phænomena*, I then endeavoured to find out the Cause of them. I was already convinced that the apparent Motion of the Stars was not owing to a Nutation [nodding motion] of the Earth's Axis. The next thing that offered itself was an Alteration in the Direction of the Plumb-line with which the Instrument was constantly rectified; but this upon trial proved insufficient. Then I considered what Refraction [by the Earth's atmosphere] might do; but here also nothing satisfactory occurred. At last I conjectured, that all the *Phænomena* hitherto mentioned, proceeded from the progressive Motion of Light and the Earth's annual Motion in its Orbit. For I perceived that, if Light was propagated in Time, the apparent Place of a fixed Object would not be the same when the Eye is at Rest, as when it is moving in any other Direction, than that of the line passing through the Eye and Object; and that, when the Eye is moving in different Directions, the apparent place of the Object would be different.

22 William Herschel on the proper motion of the sun and solar system

Bradley's observations also showed that the stars really were extraordinarily far off, and this fact became accepted by all astronomers. Moreover his work, and the prodigious star cataloguing that he carried out when he was appointed Astronomer Royal in succession to Halley, began to have its effect, most markedly on that indefatigable observer, William Herschel:

The new lights that modern observations have thrown upon several interesting parts of astronomy begin to lead us now to a subject that cannot but claim the serious attention of every one who wishes to cultivate this noble science. That several of the fixed stars have a proper motion is now already so well confirmed, that it will admit of no further doubt. From the time this was first suspected by Dr. HALLEY we have continued observations that shew Arcturus, Sirius, Aldebaran, Procyon, Castor, Rigel, Altair, and many more, to be actually in

29 *Herschel's diagrams in his paper in the* Philosophical Transactions of the Royal Society *[hereafter referred to as Phil. Trans.], vol 73, 1783 'On the proper Motion of the Sun and Solar System; . . .' (extract 22), showing the stellar motions to be expected if the Sun is travelling through space. Figure 1 (left) is the illustration from the text of the extract. Figure 2 (right) shows how the greatest parallactic shift due to the Sun's motion will be among the stars lying to the right and left of the line AB (the direction of the Sun's motion), and that stars elsewhere (as at S on the line Sa) will show a progressively smaller parallax as their proximity to the line AB increases*

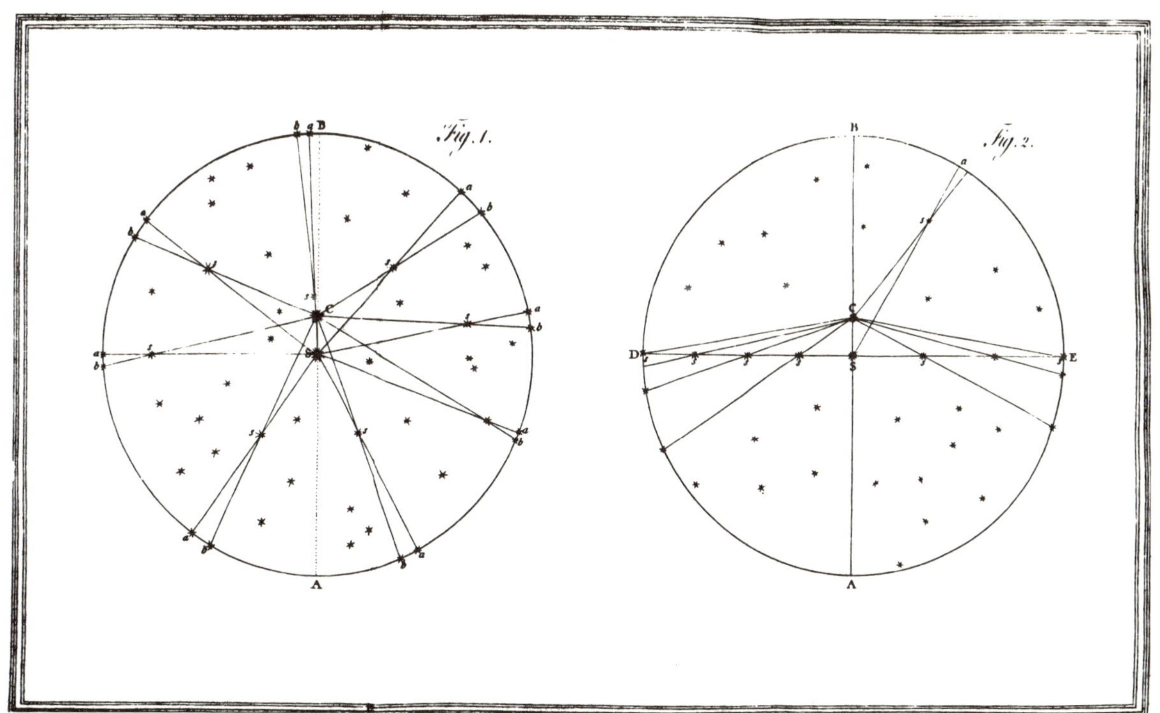

motion; and considering the shortness of the time we have had observations accurate enough for the purpose, we may rather wonder that we have already been able to find the motions of so many, than that we have not discovered the like alterations in all the rest. Besides, we are well prepared to find numbers of them apparently at rest, as, on account of their immense distance, a change of place cannot be expected to become visible to us till after many ages of careful attention and close observation, though every one of them should have a motion of the same importance with Arcturus. This consideration alone would lead us strongly to suspect, that there is not, in strictness of speaking, one *fixed* star in the heavens; but many other reasons, which I shall presently adduce, will render this so obvious, that there can hardly remain a doubt of the general motion of all the starry systems, and consequently of the solar one among the rest.

. . . . .

Suppose the sun to be at S, fig. I [figure 29]; the fixed stars to be dispersed in all possible directions and distances around at *s, s, s, s*, &c. Now, setting aside the proper motion of the stars, let us first consider what will be the consequence of a proper motion in the sun; and let it move in a direction from A towards B. Suppose it now arrived at C. Here, by a mere inspection of the figure, it will be evident, that the stars *s, s, s*, which were before seen at *a, a, a*, will now, by the motion of the sun from S to C, appear to have gone in a contrary direction, and be seen at *b, b, b*; that is to say, every star will appear more or less to have receded from the point B, in the order of the letters *ab, ab, ab*. The converse of this proposition is equally true; for if the stars should all appear to have had a retrograde motion, with respect to the point B, it is plain, on a supposition of their being at rest, the sun must have a direct motion towards the point B, to occasion all these appearances.

Herschel found that, taking the proper motions of

14 stars, he obtained a point in the sky near the star λ *(lambda) Herculis as the place to which the Sun and solar system appeared to be moving. Although his sample of 14 stars was really inadequate from a statistical point of view to give an accurate answer, later research has shown that he was surprisingly correct.*

23 William Herschel on the discovery of binary stars

But in spite of his astounding ability as an observer, Herschel was unable to obtain a precise value for any stellar parallax. He had begun noting carefully pairs of stars that appeared to be close to one another in the sky, in the hope that he could measure parallactic shifts with great accuracy. Once more, like other astronomers, he was unsuccessful. But again, like Bradley, he came upon a new discovery:

In my Remarks on the Construction of the Heavens, contained in my last Paper on this subject, I have divided the various objects which astronomy has hitherto brought to our view, into twelve classes. . . . It will not be required that I should add any thing farther on the subject of this first article of my classification; I may therefore immediately go to the second, which treats of binary sidereal systems, or real double stars.

We have already shewn the possibility that two stars, whatsoever be their relative magnitudes, may revolve, either in circles or ellipses, round their common centre of gravity; and that, among the multitude of the stars of the heavens, there should be many sufficiently near each other to occasion this mutual revolution, must also appear highly probable. But neither of these considerations can be admitted in proof of the actual existence of such binary combinations. I shall therefore now proceed to give an account of a series of observations on double stars, comprehending a period of about 25 years, which, if I am not mistaken, will go to prove, that many of them are not merely double in appearance, but must be allowed to be real binary combinations of two stars, intimately held together by the bond of mutual attraction.

24 Friedrich Bessel on the discovery of stellar parallax

Undaunted by failure, observers continued to hunt for stellar parallax, but it was not until the development of more accurate measuring equipment that success became possible. In 1829 the German astronomer Friedrich Bessel (1784-1846) took delivery of a comparatively small but beautifully made telescope by Joseph Fraunhofer (1787-1826). Mounted so that it could automatically track the stars as they move across the sky due to the Earth's rotation, it also had its front lens of $6\frac{1}{4}$ inches in diameter split in half. The purpose of this was to allow the image of one star to be superimposed on that of another by moving one of the half lenses, the amount moved being a precise measure of the angular separation of the two stars in the sky. At the Königsberg Observatory in Prussia, Bessel began a series of measurements on the star 61 Cygni [the number 61 is a cataloguing device which is an extension of Bayer's system]: he chose this star because it has a very large proper motion and so was likely to be near. In South Africa, Thomas Henderson was also making observations of a southern hemisphere star, α(alpha) Centauri, which also exhibited a huge proper motion; but he was too cautious about announcing his results, so it was Bessel who claimed priority of discovery and announcement in a letter to Sir John Herschel in October 1838:

Esteemed Sir,—Having succeeded in obtaining a long-looked for result, and presuming that it will interest so great and zealous an explorer of the heavens as yourself, I take the liberty of making a communication to you thereupon. . . . to you, I can write in my own language, and thus secure my meaning from indistinctness.

After so many unsuccessful attempts to determine the parallax of a fixed star, I though it worth while

30 *The Oxford Heliometer. This refracting telescope, with an object-glass of $7\frac{1}{2}$in diameter, was erected at Oxford University in 1848. Mounted equatorially in a similar way to Bessel's Fraunhofer heliometer refractor, and driven by 'clock-work' to follow the stars automatically, its object-glass was also cut in two halves. The heliometer was so called because the split object-glass, devised by John Dollond (1706-61), was used for measuring the diameter of the Sun. Instruments of this kind were widely used for stellar parallax determinations in the latter part of the nineteenth century. From D. Lardner and Edwin Dunkin,* Handbook of Astronomy *(London 1860). The original instrument is now in the Science Museum, London*

to try what might be accomplished by means of the accuracy which my great Fraunhofer Heliometer gives to the observations. I undertook to make this investigation upon the star 61 *Cygni*, which, by reason of its great proper motion, is perhaps best of all; Which affords the advantage of being a double star, and on that account may be observed with greater accuracy; and which is so near to the pole that, with the exception of a small part of the year, it can always be observed at night at a sufficient distance from the horizon. . . .

I selected among the small stars which surround that double star, two between 9th and 10th magnitudes; of which one (*a*) is nearly perpendicular to the line of direction of the double star; the other (*b*) nearly in this direction. I have measured with the heliometer [see figure 30] the distances of these stars from the point which bisects the distances of the two stars of 61 *Cygni*; as I considered this kind of observation the most correct that could be obtained, I have commonly repeated the observations sixteen times every night. When the atmosphere has been unusually unsteady, I have, however, made more numerous

31 *The Airy Transit Instrument at the Royal Observatory, Greenwich. Besides measuring the shifts of stars for parallax and proper (ie, real) motions, the fundamental positions of some stars must also be determined. The Royal Observatory at Greenwich, founded in 1675, had a tradition for this precise work, as had its French counterpart, the Paris Observatory, founded in 1667. While Astronomer Royal, Airy installed a number of instruments for accurate positional determinations, of which the transit circle was the most famous. Designed by Airy with the help of the engineer Charles May in 1848, it was in use by 1851. With it the moment a star crosses the meridian can be accurately determined, while precisely graduated circles enable the star's meridian altitude to be obtained. The internationally agreed meridian of zero longitude passes through the instrument, which may still be seen at the old Greenwich observatory, now part of the National Maritime Museum. From* The Illustrated London News, *volume 77 (1880)*

repetitions; although, by this, I fear the result has not attained that precision which it would have possessed by fewer observations on more favourable nights. This unsteadiness of the atmosphere is the great obstacle which attaches to all the more delicate astronomical observations. . . .

I have, therefore, deduced a . . . result from the observations, which rests on the supposition that the parallaxes of *a* and *b* are *insensible*; or that α'' and β'' are equal. For this purpose, since both series must now be brought into connexion with one another, it was necessary to deduce the *weight* [significance] of the observations contained in the second series, the weight of those in the first series being taken as unit. I have found it $= 0.6889$; and hence the most probable value of the annual parallax of 61 *Cygni* $= 0''.3136$.

Once Bessel and Henderson had shown the way, and the size of parallax angles was known, other meticulous observations were made. But it was a long and tedious process and in the next fifty years only a further two dozen stars had their parallaxes obtained.

25 George Bond on the importance of astronomical photography

But a change was to come with the development of a new observing aid—photography—although it was to be some time before it was available for precision applications. To begin with, photographic materials were comparatively insensitive, but in 1845 the French physicists Léon Foucault (1819-68) and Hippolyte Fizeau (1819-96) obtained a daguerreotype of the Sun, and five years later the American astronomer George Phillips Bond (1825-65) successfully made one of the Moon. The amount of detail obtainable on Louis Daguerre's (1789-1851) metal plates was astounding, and Bond fully realised the potentialities of the new medium:

About seven years since (July 17th 1850) Mr. *Whipple* obtained daguerreotype impressions from the image of α *Lyrae* formed in the focus of the Great Equatorial, and subsequently from *Castor*, thus establishing a simple but not uninteresting fact—the possibility of such an achievement. On these occasions a long exposure of one or two minutes was required before the plate was acted upon by the light, and in this interval the irregularities of the Munich clock-work were so great as to destroy the symmetry of the images, while the smaller [dimmer] stars of the second magnitude would not take at all.

For some years after Mr. *Whipple* gave his attention to photographs of the moon and sun, and the stars were left to themselves. But improvements in the art progressed rapidly; the preparations [photographic emulsions] were more sensitive; the artists had acquired more experience. At the same time the principle of the spring governor had been thoroughly tested, and found to supply a great desideratum in imparting a sidereal motion to the telescope, incomparably more uniform than that attained by the Munich mechanism. Messrs. *Whipple* and *Black* recommended their trials on star images (taken by the collodion process) in March this year, and they are still in progress. . . . But the results already obtained in the disconnected attempts we have thus far been enabled to make are of the highest interest, and suggest possibilities in the future which one can scarcely trust himself to speculate upon. Could another step in advance be taken equal to that gained since 1850, the consequences could not fail of being of incalculable importance in Astronomy. . . .

On a fine night the amount of work which can be accomplished with entire exemption from the trouble, vexation and fatigue that seldom fail to attend upon ordinary observations, is astonishing. The plates, once secured, can be laid by for future study by daylight and at leisure. The record is there, with no room for doubt or mistake as to its fidelity. As yet, however, we obtain images only from stars to the sixth magnitude inclusive. To be of essential service

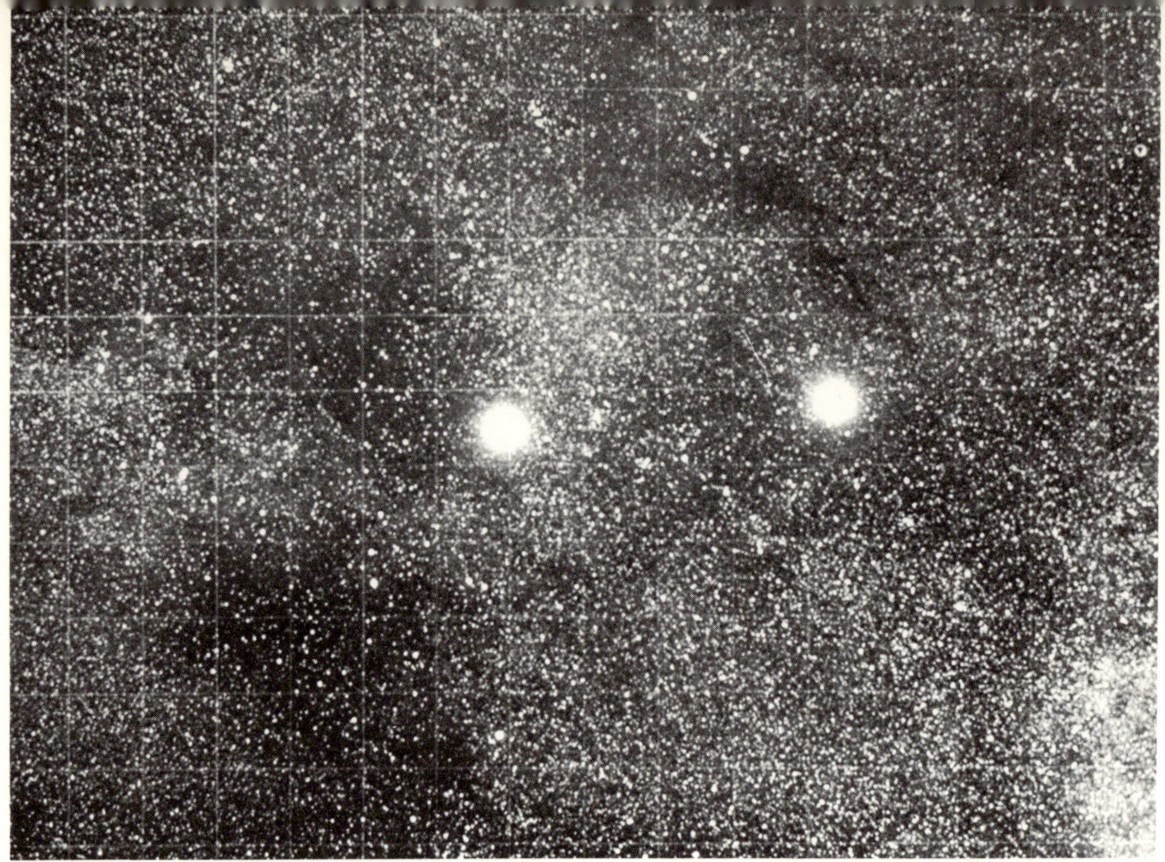

32 *In 1885 astronomers from seventeen different nations met in Paris and agreed to co-operate in a photographic survey of the sky. Known as the* Carte du Ciel, *the scheme was never completed, although a great number of the required 20,000 photographs were taken. This photograph, taken at Johannesburg in 1910 by Franklin Adams, an English amateur astronomer, shows the Milky Way in the region of α and β Centauri. Notice the rectangular grid lines across the photograph, placed there to assist in measuring the positions of the stars. This was a plate from a complete survey that Franklin Adams made and that was directly inspired by the* Carte du Ciel *project*

to Astronomy, it is indispensable that great improvements be yet made, and these I feel sure will not be accomplished without a deal of experimenting.

. . . . . .

What more admirable method can be imagined for the study of the orbits of the fixed stars and for resolving the problem of their annual parallax than this would be, if we could obtain the impressions of the telescopic stars to the tenth magnitude? Consider, that groups of ten, or fifty even, if so many occur in the compass of the field, will be taken as quickly as one alone would be—perhaps in a few seconds only—and each mapped out with unimpeachable accuracy.

I have not alluded to two important features in stellar photography. One is that the intensity and size of the images, taken in connection with the length of time during which the plate has been exposed, measures the relative magnitudes of the stars. The other point is that the measurements of distances and angles of position of the double stars from the plates we have ascertained, by many trials on our earliest impressions, to be as exact as the best micrometric work. Our subsequent pictures are much more perfect, and should do better still.

Once photographic plates had been invented and more sensitive photographic emulsions to go with them, photography became an integral part of astronomy. For positional measurement, the plate could be placed in a measuring microscope and a degree of precision obtained that was impossible with just the eye and the telescope.

Thus by 1924 the number of stars whose parallax had been measured was 1,600, and the angles that could be measured had dropped to as little as one fiftieth of a second of arc.

26 Spencer Jones on the solar parallax

All stellar parallax measures depend upon the value for the size of the Earth's orbit, and this means determining the distance between Earth and Sun as precisely as possible. In the seventeenth century a number of attempts were made to determine this and values of the order of 87 million miles were obtained—the first measures with any pretensions to accuracy. The method used was based on Kepler's third law of planetary motion which, if one measures the distance from the Earth to a nearby planet, allows one to calculate one's distance from the Sun. Halley strongly advocated observing Venus when it appears to transit across the Sun's disk, because then it comes closer to the Earth than Mars, the only other nearby planet. Such transits are rare, but so successful was Halley's advocacy that after his death, the transits of 1761 and 1769 were observed from all over the world. A later analysis of these observations gave 8".5 for the Sun's parallax, equivalent to a distance of $95\frac{1}{4}$ million miles.

So fundamental a value has been a perpetual challenge to each generation of astronomers, and other, later, determinations have been made. With the discovery of the planetoid Eros, which has an eccentric orbit lying between those of Mars and the Earth, fresh stimulus was given to the problem in the twentieth century. The following is an extract from Spencer Jones' address to the Royal Astronomical Society on the determination of the solar parallax:

1. At the opposition of 1931, the asteroid [planetoid] 433 Eros approached the Earth to a

33 *Plate measuring machine at the Cambridge Observatories, University of Cambridge*

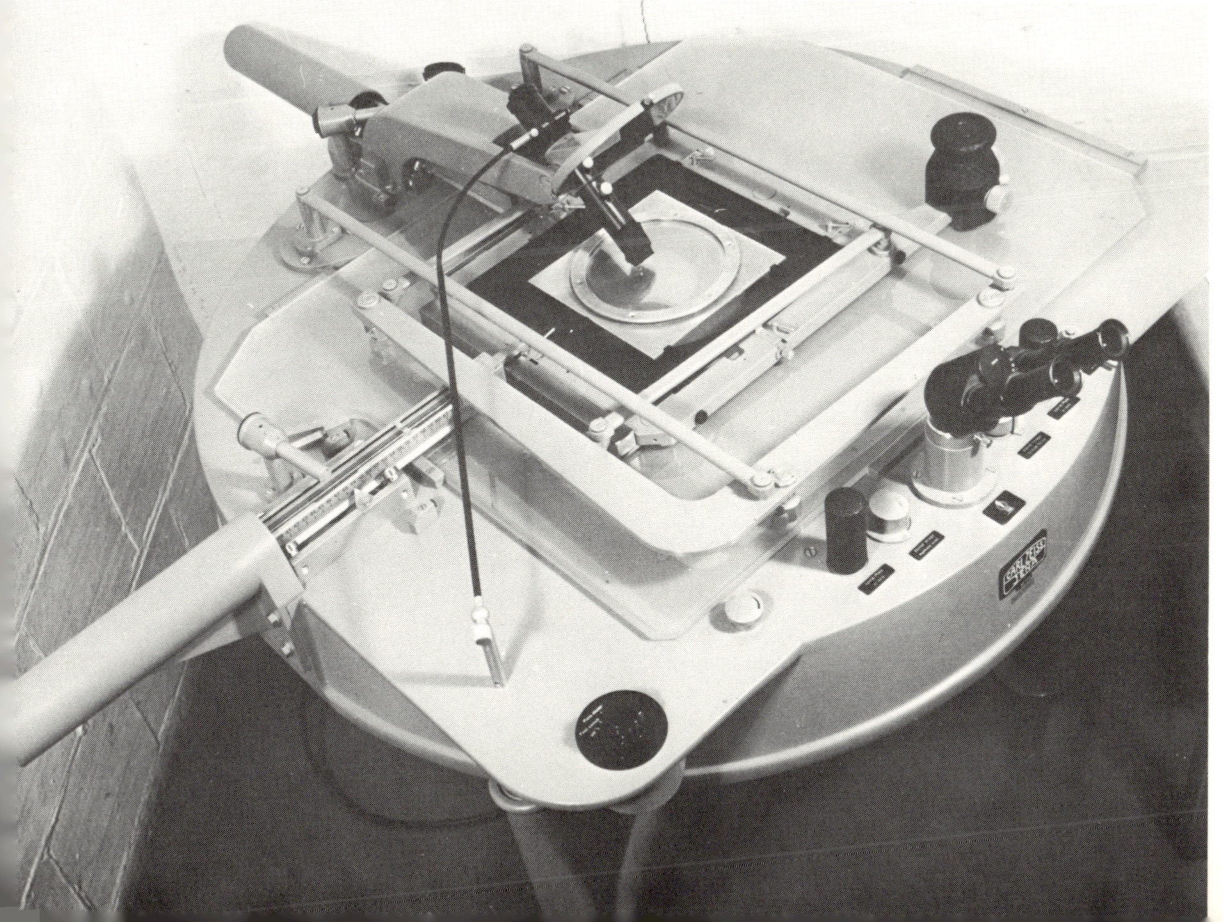

34 *The General Automatic Luminosity and XY automatic measuring machine (GALAXY). Using special positional and detection techniques to find and measure star images, utilizing moiré fringes and 'flying spot' scanners, this electronically controlled machine has automated much hitherto laborious and time-consuming work on the examination of photographic plates. It was designed jointly by Dr V. C. Reddish and colleagues at the Royal Observatory, Edinburgh, and the manufacturers, Faul Coradi Limited. The photographic plates are inserted into the holder that lies below the vertical cylindrical units seen through the open door. The instrument uses plates, taken with the Observatory's 16-inch Schmidt type of telescope (see also figure 84) and their Elliott 4130 computer*

distance of 16.2 million miles, being its closest approach since its discovery in 1898. This provided a particularly favourable occasion for the determination of the solar parallax, one of the fundamental constants in astronomy. The path of Eros in the sky near opposition formed a large arc, stretching from declination $+44°$ to $-26°$, enabling observations to be obtained at observatories in both hemispheres. Independent determinations of the solar parallax thus

became possible from the observations in right ascension and from those in declination; in right ascension, the solar parallax could be determined from the comparison between observations taken at easterly and westerly hour-angles with the same instrument; in declination, it could be determined from comparisons between observations made practically simultaneously at two observatories, one in the northern and the other in the southern hemisphere.

2. At the General Assembly of the International Astronomical Union in 1928, a Solar Parallax Commission was appointed, with the writer as President, to make arrangements for the co-operative effort to secure the observations and for their subsequent reduction and discussion. . . .

SUMMARY

1. The observations of Eros, obtained around the opposition of 1931, have been discussed and determinations of the solar parallax and of the constant of the lunar equation have been made.

2. The solar parallax is $8''.790 \pm 0''.001$. Independent determinations from R.A. [Right Ascension] and Dec. [Declination] observations are in close agreement. The results have a high internal consistency.

3. The discussion of possible systematic error arising from differential atmospheric dispersion shows that atmospheric dispersion effects are bound up with the chromatic properties of the objectives and other factors. The effects of colour differences between Eros and the reference stars tend in consequence to assume a random nature. An investigation of the mean effect, which should depend on the distance of Eros, suggests that it is very small.

Spencer Jones' value gave the Sun's distance as 92,910,300 miles, but more recent determinations have been obtained both from observations of Eros and also by other means, and the best present value is $8''.79405$ or 92,957,219 miles.

CHAPTER THREE

The Nature of the Stars

27 Aristotle on the stars

To the Greeks the stars were eternal. They had been observed by countless generations and civilisations before their own, and never suffered any apparent change. Thus, Aristotle could use the concept of the quintessence, the changeless fifth element, in his description of the stars:

It would be most natural and consequent upon what has been said that each of the stars [these include the Sun, Moon and planets] should be composed of that substance in which their path lies [ie of the same substance as the spheres], since, as we said, there is an element whose natural movement is circular [the fifth element or 'quintessence' which comes no lower than the sphere of the Moon]. In so saying we are only following the same line of thought as those who say that the stars are fiery because they believe the upper body to be fire, the presumption being that a thing is composed of the same stuff as that in which it is situated. The warmth and light which proceed from them are caused by the friction set up in the air by their motion. Movement tends to create fire in wood, stone, and iron; and even with more reason should it have that effect on air, a substance which is closer to the fire than these. An example is that missiles, which as they move are themselves fired so strongly that leaden balls are melted; and if they are fired the surrounding air must be similarly affected. Now while the missiles are heated by reason of their motion in the air, which is turned into fire by the agitation produced by their movement, the upper bodies are carried on a moving sphere, so that, though not themselves fired, yet the air underneath the sphere of the revolving body is necessarily heated by its motion, and in particular in that part where the Sun is attached to it. Hence warmth increases as the Sun gets nearer or higher or overhead. Of the fact, then, that the stars are neither fiery nor move in fire, enough has been said.

And there seemed little more that could be said, for

the stars were too far off for any contrary evidence to be found.

28 Tycho Brahe on the Nova of 1572

However, in 1572 a nova or 'new' star suddenly appeared in the constellation of Cassiopeia, and was observed all over western Europe, causing much comment. In fact, it was so bright that it was what now would be called a 'supernova'. Among those who saw it was Tycho Brahe who, the very next year, published a pamphlet about it in Copenhagen:

On the eleventh of November of last year [1572], in the evening after sunset, I was gazing at the stars in a clear sky, as is my custom, when I noticed a new and extraordinary star, brighter than all the others shining almost directly overhead; and since, almost from boyhood, I had known perfectly every star in the sky (there is no great difficulty in gaining this knowledge), it was obvious to me that there never had been a star at that place in the sky, not even the smallest, to say nothing of one so conspicuously bright as this: it was so astonishing that there was no shame in doubting the trustworthiness of my own eyes. Yet when others saw that there really was a star there, when the place was pointed out to them, I had no more room for doubt. . . .

How accordingly it might become known to us by calculation, in which region of the Elements or among which celestial orbs this star might be, . . . we searched out whether it had any parallax and how much, . . . Then I observed numerous other stars in a similar way wherefore I conclude that this new star has no difference in appearance, even when close to the horizon. . . . Therefore it is necessary to place this star not in the region of the Elements below the Moon, but a long way above in a sphere with respect to which the earth has no sensible size.

This was an epoch-making discovery, for it meant

that changes did occur in the heavens, and were not confined to the region below the sphere of the Moon. It was the first sound observational evidence that all was not well with the Aristotelian universe. And in 1577 when a very bright comet appeared, similar measurements by Tycho, coupled with those of others, made it clear that this body, like the nova, lay beyond the sphere of the Moon. As it moved through the sky, it must cut right through the celestial spheres and, once again, the old concepts were shown to be in error.

29 William Herschel on the nature of the Sun

In England the whole outlook on the universe was changing rapidly, and in 1576 Thomas Digges, whose observations of the nova Tycho had incorporated with his own, suggested that the stars were bodies like the Sun, although the nature of the Sun itself was uncertain. When the telescope arrived, much time was spent observing the Sun: changing spots were seen on its surface, and no longer could it be thought of as a 'perfect' body as Aristotle had taught. But little other evidence was available, and even as late as 1795 William Herschel could make suggestions about solar inhabitants:

. . . That the sun has a very extensive atmosphere cannot be doubted; and that this atmosphere consists of various elastic fluids, that are more or less lucid and transparent, and of which the lucid one is that which furnishes us with light, seems also to be fully established by all the phænomena of its spots, of the faculæ [very bright patches], and of the lucid surface itself. There is no kind of variety in these appearances but what may be accounted for with the greatest facility, from the continual agitation which we may easily conceive must take place in the regions of such extensive elastic fluids.

It will be necessary, however, to be a little more particular, as to the manner in which I suppose the lucid fluid of the sun to be generated in its atmosphere. An analogy that may be drawn from the generation of clouds in our own atmosphere, seems to be a very proper one, and full of instruction. Our clouds are probably decompositions of some of the elastic fluids of the atmosphere itself, when such natural causes, as in this grand chemical laboratory are generally at work, act upon them; we may therefore admit that in the very extensive atmosphere of the sun, from causes of the same nature, similar phænomena will take place; but with this difference, that the continual and very extensive decompositions of the elastic fluids of the sun are of a phosphoric nature, and attended with lucid appearances, by giving out light.

. . . .

Before I proceed, I shall only point out that . . . a dark spot in the sun is a place in its atmosphere which happens to be free from luminous decompositions; and that faculæ are, on the contrary, more copious mixtures of such fluids as decompose each

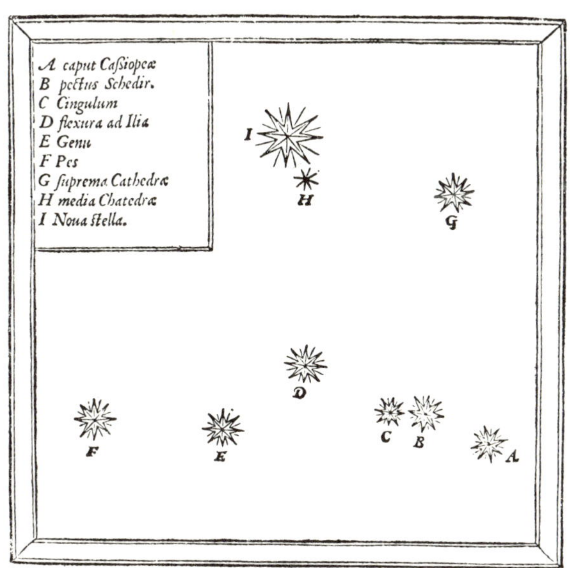

36 *The nova of 1572 in* Cassiopaeia, *as shown by Tycho in his* De Stella Nova (On the New Star) *published in Copenhagen the following year*

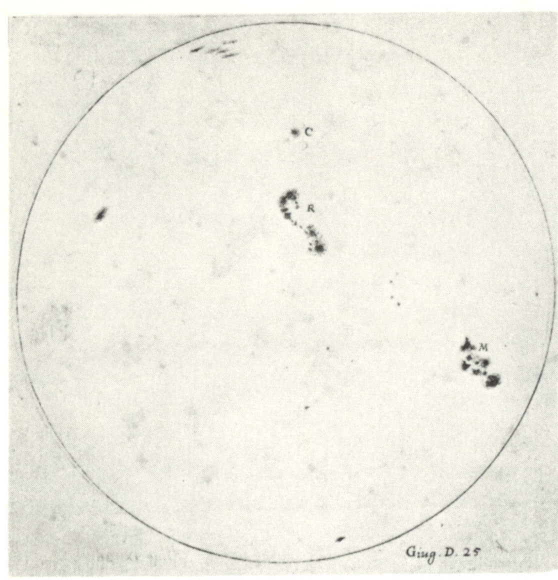

37 *Drawing of sunspots by Galileo and published in his* Istoria e Dimonstrazioni Intorno Alle Macchie Solari *(History and Demonstrations Concerning Spots on the Sun) (Rome 1613)*

other. The penumbra which attends spots, being generally depressed more or less to about half way between the solid body of the sun and the upper part of those regions in which luminous decompositions take place, must of course be fainter than other parts. . . .

From the luminous atmosphere of the sun I proceed to its opaque body, which by calculation from the power it exerts upon the planets we know to be of great solidity; and from the phænomena of the dark spots, many of which, probably on account of their high situations, have been repeatedly seen, and otherwise denote inequalities in their level, we surmise that its surface is diversified with mountains and vallies. . . .

The sun, viewed in this light, appears to be nothing else than a very eminent, large, and lucid planet,

evidently the first, or in strictness of speaking, the only primary one of our system; all others being truly secondary to it. Its similarity to the other globes of the solar system with regard to its solidity, its atmosphere, and its diversified surface; the rotation upon its axis, the fall of heavy bodies, leads us on to suppose that it is most probably also inhabited, like the rest of the planets, by beings whose organs are adapted to the peculiar circumstances of that vast globe.

30 Kirchoff on the solar spectrum

Yet within the next 75 years the picture was to change remarkably. This was due to the analysis of sunlight, which Newton had begun in 1665 by spreading the light out into a coloured spectrum to help his investigation of light and colours. To study the question further, in 1802 William Wollaston passed light through a narrow slit, and found that the coloured spectrum was crossed by many dark lines. His discovery stimulated others to try to find what caused the lines because one thing was clear—whatever they might be, they were not gaps between the colours of the spectrum. Joseph Fraunhofer mapped a great number of them (figure 39). But it was only after laboratory experiments were added to solar observations that a solution was found. In the laboratory when the chemical element sodium was heated it emitted a pair of yellow lines when observed with a spectroscope (figure 40)—known from Fraunhofer's reference letters as the D lines—and these coincided with a pair of dark lines in the yellow part of the spectrum. In Berlin, Gustav Kirchhoff (1824-87) and Robert Bunsen (1811-99) followed up this clue. They found coincidences between dark lines in the solar spectrum and in the spectrum of vaporised iron, and Kirchhoff saw clearly the explanation:

Hence this coincidence must be produced by some cause; and a cause can be assigned which affords a perfect explanation of the phenomenon. The observed phenomenon may be explained by the supposition

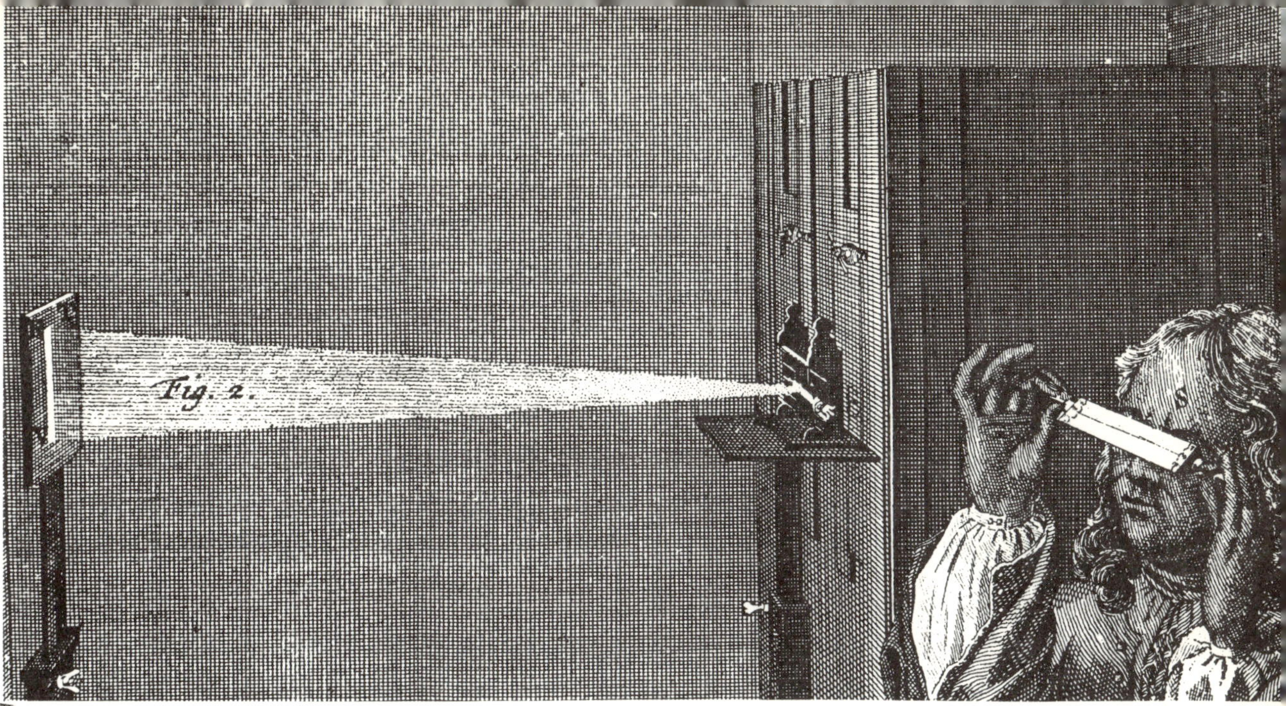

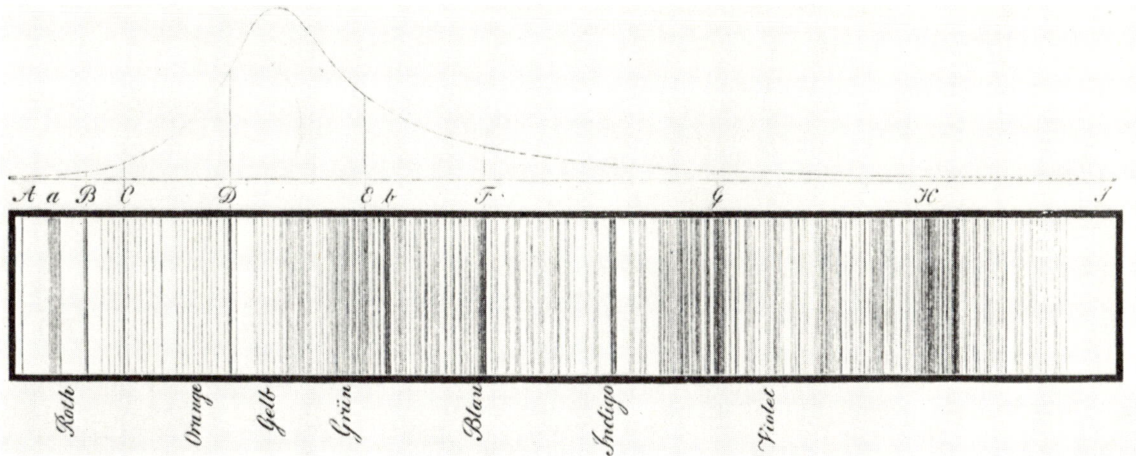

38 *Newton's analysis of sunlight, in which he demonstrated that 'white' light is composed of light of all colours. From John Desaguliers,* Mathematical Elements of Natural Philosophy, confirm'd by Experiments *(London 1747). This was an English translation of* Physices Elementa Mathematica *(Leyden 1720) by Newton's supporter Willem s'Gravesande (1688-1742)*

that the rays of light which form the solar spectrum have passed through the vapour of iron, and have thus suffered the absorption which the vapour of iron must exert. . . .

As soon as the presence of *one* terrestrial element in the solar atmosphere was thus determined, and thereby the existence of a large number of Fraunhofer's lines explained, it seemed reasonable to suppose that other terrestrial bodies occur there, and that, by exerting their absorptive power, they may cause the production of other Fraunhofer's lines. For it is very probable that elementary bodies which

39 *Fraunhofer's drawing of the lines of the solar spectrum with, above it, a curve representing the intensity of sunlight in different parts of the spectrum. From H. Roscoe,* Spectrum Analysis *(London 1870)*

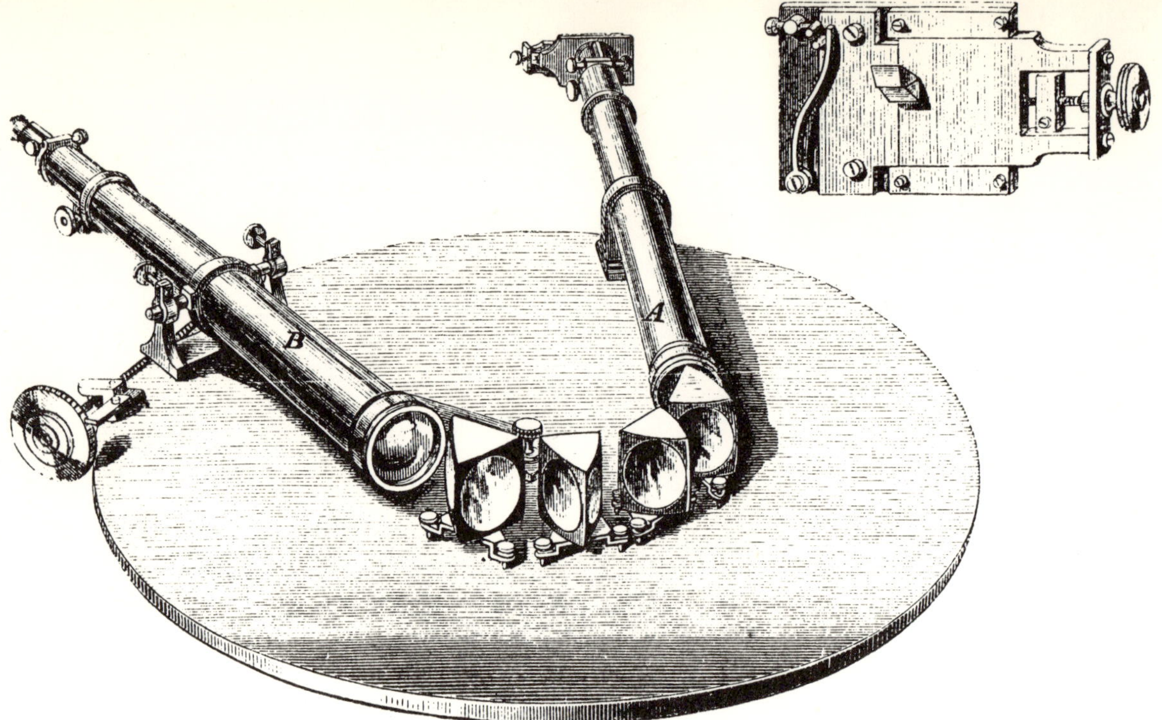

40 *The spectroscope used by Kirchhoff to analyse the solar spectrum, and constructed for him by C. A. Steinheil of Munich. The inset (upper right) shows the slit to which a small right-angled prism is fixed so that the solar spectrum (which falls on the rest of the slit) may be compared with radiation from a laboratory source (which is caused to fall on the prism). From the same source as figure 39*

occur in large quantities on the earth, and are likewise distinguished by special bright lines in their spectra, will, like iron, be visible in the solar atmosphere. This is found to be the case with calcium, magnesium, and sodium. . . .

31 Huggins and Miller on stellar spectra

At last astronomers had a means of finding out something that had hitherto appeared impossible—the chemical nature of the Sun and, by implication, of the stars as well. But to obtain a solar spectrum is one thing, to obtain a stellar spectrum is a different matter, for even the brightest star appears some 25 thousand million times dimmer than the Sun. All the same, attempts were made and William Huggins (1824-1910) and William Miller (1817-70), working at Huggins' private observatory in south London, devised new techniques that were to prove invaluable:

The investigation of the nature of the fixed stars by a prismatic analysis of the light which comes to us from them, however, is surrounded with no ordinary difficulties. The light of the bright stars, even when concentrated by an object-glass or speculum [mirror] is found to become feeble when subjected to the large amount of dispersion which is necessary to give certainty and value to the comparison of the dark lines of the stellar spectra with the bright lines of terrestrial matter. Another difficulty, greater because it is in its effect upon observation more injurious, and is altogether beyond the control of the experimentalist, presents itself in the ever-changing want of homogeneity of the earth's atmosphere through which the stellar light has to pass. This source of difficulty presses very heavily upon observers who have to work in a climate so unfavourable in this respect as our own. On any but the finest nights the numerous and closely approximated fine lines of the stellar spectra are seen so fitfully that no observations of value can be made. It is from this cause especially that we have found the inquiry, in which for more than two years and a quarter we have been engaged, more than usually toilsome; and indeed it has demanded a sacrifice of time very great when com-

58

pared with the amount of information which we have been enabled to obtain. . . .

Early in 1862 we had succeeded in arranging a form of apparatus in which a few of the stronger lines in some of the brighter stars could be seen. The remeasuring of those already described by Fraunhofer and Donati, and even the determining the positions of a few similar lines in other stars, however, would have been of little value for our special object, which was to ascertain, if possible, the constituent elements of the different stars. We therefore devoted considerable time and attention to the perfecting of an apparatus which should possess sufficient dispersive and defining power to resolve such lines as D and *b* of the solar spectrum. Such an instrument would bring out the finer lines of the spectra of the stars, if in this respect they resembled the sun. It was necessary for our purpose that the apparatus should further be adapted to give accurate measures of the lines which should be observed, and that it should also be so constructed as to permit the spectra of the chemical elements to be observed in the instrument simultaneously with the spectra of the stars. In addition to this, it was needful that these two spectra should occupy a position relatively to each other as to enable the observer to determine with certainty the coincidence or non-coincidence of the bright lines of the elements with the dark lines in the light from the star.

Before the end of the year 1862 we had succeeded in constructing an apparatus which fulfilled part of these conditions. . . .

The technique of observing laboratory spectra next to stellar spectra was of the most vital importance if precision was to be brought to the question of identification of chemical elements, and, as it turned out later, to determining the physical conditions on the stars: it was also to have other implications, as will become evident, and is the standard technique today.

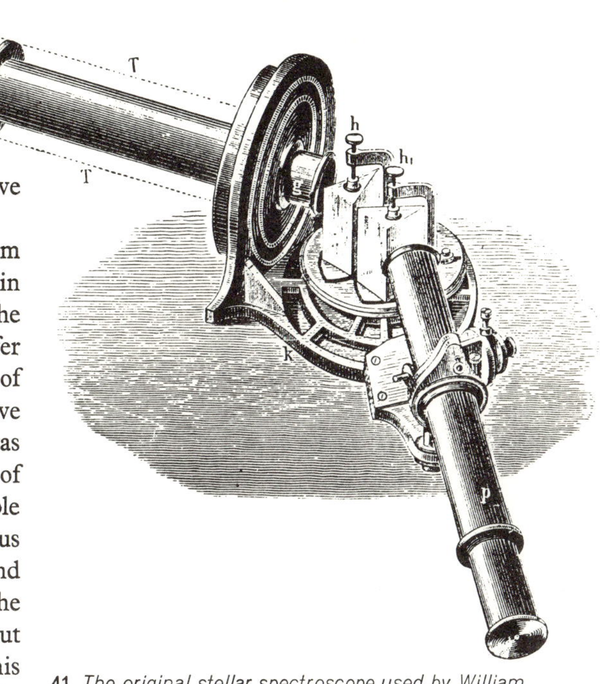

41 *The original stellar spectroscope used by William Huggins on his 8-inch diameter refractor at his observatory at Tulse Hill, and constructed for him by Browning of London. The telescope eyepiece tube was at T, and inside this slid another tube that carried the cylindrical lens A which spread the stellar image over the slit of the spectroscope c, which carries a small right-angled prism for providing comparison spectra. A collimating lens g causes a parallel beam of starlight to pass to the prisms h, h mounted on the casting k. The spectrum was observed through the viewing telescope p, whose position relative to the prisms could be adjusted by the micrometer screw, so that different parts of the spectrum could be examined in detail. From H. Schellen,* Spectrum Analysis, *translated by Jane and Caroline Lassell and edited with notes by William Huggins (London 1872)*

32 Janssen on observing prominences

The Sun, it was realised, was the closest star to us, and some astronomers concentrated their attention on it, notably Norman Lockyer (1836-1920) and Pierre Janssen (1824-1908). One problem that arose was the appearance of the Sun and Moon during a total solar eclipse, for huge, bright pink flame-like 'prominences' appeared around the edge of the Sun and, in addition, a dim pearly coloured light—the corona—extended out

42 *Lockyer's solar spectroscope. Sunlight was fed to the large stationary instrument by a clockwork driven mirror outside the laboratory. From J. Norman Lockyer,* Studies in Spectrum Analysis *(London 1904)*

far into space, a light which might possibly be an appendage of the Moon, not the Sun. The trouble was that neither could ordinarily be seen: they were visible only during the few minutes of a total solar eclipse when the Moon blotted out the bright disk of the Sun. It was during the total eclipse of 1868 that Lockyer succeeded in obtaining a spectrum of a prominence, while Janssen also succeeded in observing a spectrum, and in addition, chancing upon a technique for observing prominences even when there was no total eclipse:

Immediately after totality began, two magnificent prominences appeared: one of them, more than three minutes in height [that is almost a tenth of the Sun's diameter], shining with a brilliance that is hard to imagine. Analysis of the light immediately showed me that it was formed by an immense incandescent gaseous column, chiefly composed of hydrogen gas....

But the most important result of these observations is the discovery of a method, of which the principle was contrived during the eclipse itself, which allows the study of prominences and of the circumsolar regions at all times, without the need of having to interpose an opaque body in front of the sun's disk. This method is based on the spectral properties of the light of prominences, light which resolves into a small number of very bright pencils of rays, corresponding to some of the dark lines of the solar spectrum.

No later than the day following the eclipse, the method was applied with success, and I have been able to witness phenomena presented by a new kind of eclipse which lasted the whole day long. The prominences of the day before were profoundly modified. Hardly any trace remained of the large prominence, and the distribution of the gaseous material was quite different.

Huggins improved upon the technique, but the corona remained something of a mystery. The spectrum of its inner part was examined by Janssen at the total eclipse of 1871, and he concluded that it really was part of the circumsolar regions. Yet not until 1930 was it possible to observe the spectral lines of the corona without the need of a total eclipse, when the French astronomer Bernard Lyot (1897-1952) developed the 'coronagraph';

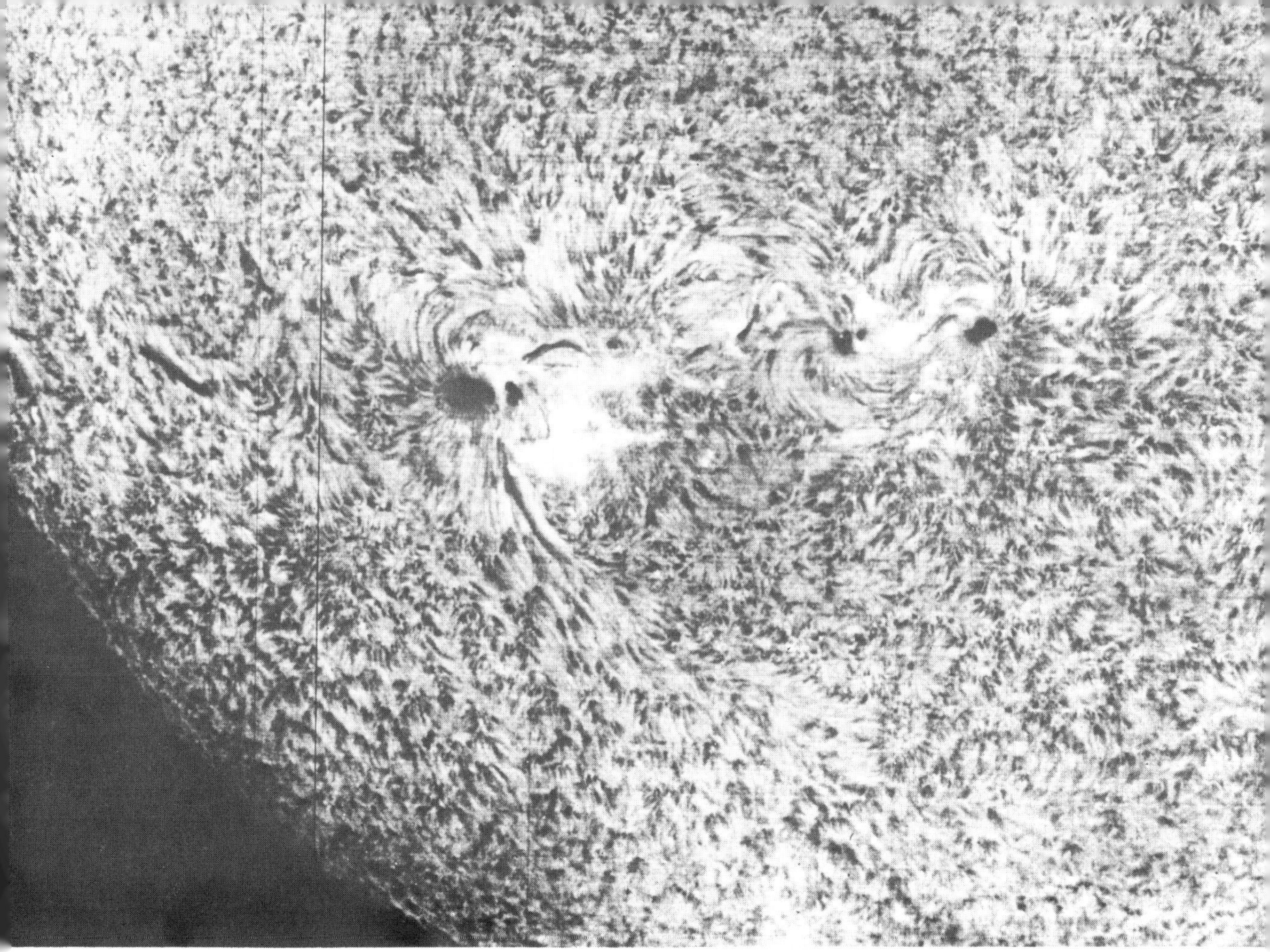

43 *Spectroheliogram of the Sun's disk photographed in the light of hydrogen (Hα) by George E. Hale (1868-1938). The bright patches Hale called 'flocculi'. The spectrohelioscope (called a spectroheliograph when used for taking photographs—spectroheliograms) was Hale's invention and is, in essence, a spectroscope with two slits, the first in the usual place in front of the prisms, the second in the plane of the spectrum and positioned to isolate the light of a particular element. The slits then scan the solar image. This photograph was taken using a spectroheliograph mounted below ground with sunlight fed from a mechanically driven mirror mounted 60ft above the ground at Mount Wilson Observatory in 1917*

44 *Solar prominence photographed through a spectroheliograph, Kodaikanal Observatory, India by John Evershed, on 1916 May 26*

Fig. 4.—Temperature curve, showing the relative temperatures of the different orders of celestial bodies. The top of the curve represents the highest temperatures, and the bottom of each arm the lowest. On the left arm, the temperatures are increasing, on the right they are decreasing. The diagram shows the relative temperatures of Vogel's classes.

45 *Lockyer's classification of the spectra of all classes of celestial object by temperature. From Lockyer's Bakerian Lecture of 1888 to the Royal Society, London, and titled 'Suggestions on the Classification of the various Species of Heavenly Bodies. A Report to the Solar Physics Committee'. Published in* Proceedings of the Royal Society, *volume 44 (1888)*

and it was not until 1941 that the Swedish physicist Bengt Edlen found that the spectral lines which had defied detection for almost 70 years were caused by extraordinarily heated atoms of calcium, iron and nickel. But the fact that the Sun was a hot glowing body, with a temperature close to its surface gases of 6,000° Centigrade (Celsius) had been known since the 1890s, a temperature that made William Herschel's belief in an inhabited Sun impossible to countenance any longer.

33 Annie Cannon on stellar classification

It became clear that the differences in the spectra of stars was one of temperature and Lockyer produced a temperature scale for all celestial bodies (figure 45), classifying spectra into temperature groupings. But it was Henry Draper (1837-82) at Harvard College Observatory who, with a team of brilliant assistants to follow him, developed the classification of Lockyer and the Italian spectroscopist Fr Angelo Secchi (1818-78), and laid the foundations of the scheme now used, a scheme that has proved a vital help to our understanding of the nature of the stars:

As soon as the photographs [of a series of spectra] were obtained, the work of classifying the spectra was undertaken, and was soon placed under the charge of Mrs. Fleming. At once the greatest

62

difficulty was encountered. How were the various kinds of stellar spectra shown on these photographs to be designated? The divisions into five types made by Secchi proved to be altogether inadequate to represent the numerous differences seen on the photographs. A new system had to be adopted which would permit the reader to understand the various aspects of the spectra as shown by the photographs. Therefore, the letters of the alphabet from A to Q were assigned to stellar spectra. This classification is purely empirical, being based wholly on the external appearances, without any idea of expressing differences of temperature or stages of evolution. . . . The futility of attempting more at this early epoch is shown by the passing of Vogel's classification, in which the aim was made to explain the phase of development of each star. In the Draper classification, the letter A was assigned to spectra of [Secchi's] first type, showing the broad hydrogen lines, as in Sirius, the line K of calcium also generally being present. When other lines were seen, such as those at wavelengths 4026 and 4471, the spectra were called B. The letters C and D were used to represent spectra of the first type, having certain peculiarities, such as double lines or bright bands, which were even then suspected to be instrumental rather than real. The letters E to L were assigned to spectra assumed to be of the second type, with the remark that Class F might be considered to be intermediate between the first and second types. The letters M, N, O, and P were given, respectively, to spectra of the third type, the fourth type, the fifth type, consisting mainly of bright lines, and the gaseous nebulae. Q was left for spectra so peculiar as not to be included under any of the former letters. The first classification of a large number of photographic stellar spectra was made according to this system by Mrs. Fleming. . . .

The appearances for which some of the letters, such as C, D and E, had been assigned, were not confirmed by later and better photographs. Therefore,

these letters were dropped from the sequence. In 1891, Professor Pickering wrote, "The principal question now outstanding is to determine what substance or substances cause the characteristic lines in the spectra of stars of the Orion type." This question was settled by Sir William Ramsay's discovery of helium in 1895, and the subsequent identification by Vogel of the lines characteristic of spectra of the Orion type with the new terrestrial element. Hence the so-called Orion stars, which were first known to prevail in that constellation, became helium stars. As it had been clearly proved by the Harvard Classification that these spectra precede Sirian spectra, it was necessary to place the letter B, which had been assigned to the Orion stars, before the letter A, or to change all the stars previously lettered A and B. Since several thousand had already been published, the change of the order of the letters was the only practicable course. This inversion of letters is variously regarded by astronomers as an advantage or a drawback to the system. The original letters that persisted were B, A, F, G, K, M, to represent the sequence as far as it was then established. . . . But, as was found in classifying the bright southern stars, the letter B could not stand for all the helium stars with their various intensities of lines and differences in number of lines present. Therefore, the writer adopted the plan of dividing into tenths the intervals between spectra represented by successive letters in the sequence. Thus the various subdivisions of B stars were called B_1A, B_2A, B_3A, B_5A, B_8A, and B_9A, later abbreviated to B_1, B_2, etc.

By the time the new Harvard classification was ready, more stellar distances were known, and following on the work of the Dane Ejnar Hertzsprung (1873-1967) who examined proper motions to obtain an idea of distance, and the spectral classification of stars, Henry Norris Russell (1877-1957) studied the relationship between absolute magnitude [a measure of the real brightness of a star] and spectral class. This work, taking

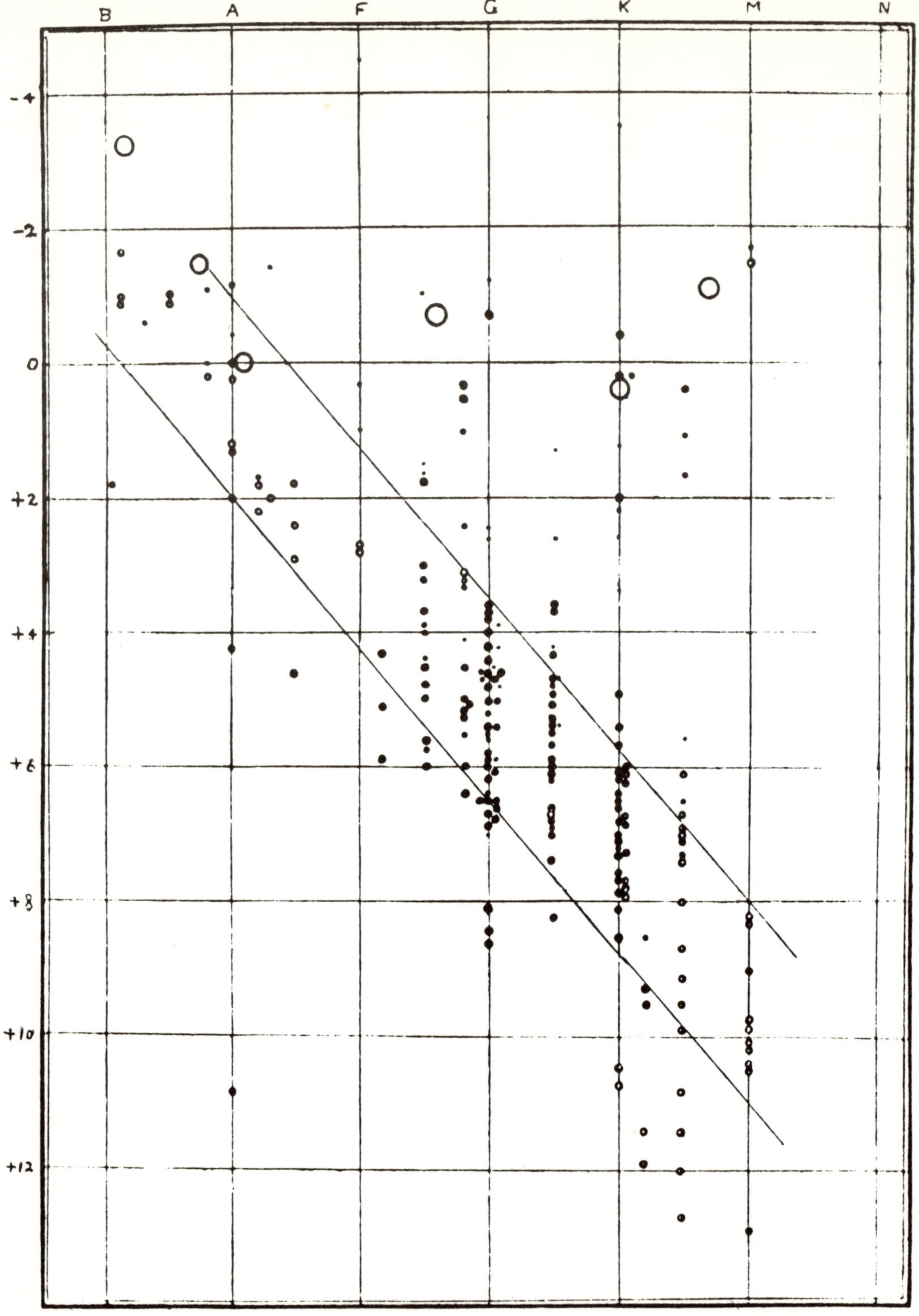

the latest results up to 1913, *culminated in the famous Hertzsprung-Russell* (HR) *diagram (figure* 46). *This diagram could be, and was, used to help obtain clues about a star's life cycle; it could also be used to determine some stellar distances, since if a star's spectrum could be classified, then its absolute magnitude would be known and, by comparing the latter with the magnitude it appeared in the sky, the distance could be computed.*

34 Helmholtz on the Sun's store of force

The generation of stellar energy also began to concern astronomers and physicists. The new outlook that came into geology in the early nineteenth century, with a long time-scale for the age of the Earth, made it necessary to find some other explanation than burning to account for the Sun's supply of energy. The well-known nineteenth-century physicist Hermann von Helmholtz (1821-94) *discussed the matter, first disposing of chemical action, and then considering other forces:*

On earth the processes of combustion are the most abundant source of heat. Does the sun's heat originate in a process of this kind? To this question we can reply with a complete and decided negative, for we now know that the sun contains terrestrial elements with which we are acquainted. . . . Calculation shows that . . . the heat resulting from their combustion would be sufficient to keep up the radiation of heat from the sun for 3,021 years. That, it is true, is a long time, but even profane history teaches that the sun has lighted and warmed us for 3,000 years, and geology puts it beyond doubt that this period must be extended to millions of years.

.   .   .   .   .

We must seek for forces of far greater magnitude, and these we can only find in cosmical attraction. We have already seen that the comparatively small masses

46 *The Hertzsprung-Russell diagram in its original form, as it appeared in* Nature, *volume 93, in the issue of 1914 May 14*

of shooting-stars and meteorites can produce extraordinarily large amounts of heat when their cosmical velocities are arrested by our atmosphere. . . .

Now we may assume with great probability that very many more meteors fall upon the sun than upon the earth, and with greater velocity, too, and therefore give more heat. Yet the hypothesis, that the entire amount of the sun's heat which is continually lost by radiation, is made up by the fall of meteors, a hypothesis which was propounded by Mayer, and has been favourably adopted by several other physicists, is open, according to Sir W. Thomson's [Lord Kelvin] investigations, to objection; for, assuming it to hold, the mass of the sun should increase so rapidly that the consequences would have shown themselves in the accelerated motion of the planets. The entire loss of heat from the sun cannot at all events be produced in this way; at the most a portion, which, however, may not be inconsiderable.

If, now, there is no present manifestation of force sufficient to cover the expenditure of the sun's heat, the sun must originally have had a store of heat which it gradually gives out. But whence is this store? We know that the cosmical forces alone could have produced it. And here the hypothesis, . . . as to the origin of the sun, comes to our aid. If the mass of the sun had been once diffused in cosmical space, and had then been condensed—that is, had fallen together under the influence of celestial gravity—if then the resultant motion had been destroyed by friction and impact, with the production of heat, the new world produced by such condensation must have acquired a store of heat not only of considerable, but even colossal, magnitude.

.   .   .   .   .

We may, . . . assume with great probability that the sun will still continue in its condensation, even if it only attained the density of the earth—though it will probably become far denser in the interior owing to the enormous pressure—this would develop fresh quantities of heat, which would be sufficient to maintain for an additional 17,000,000 of years the same intensity of sunshine as that which is now the source of all terrestrial life.

But this ingenious proposal was supplanted in the twentieth century. Theoretical calculations were made of the conditions inside the Sun and other stars, and it became clear that the central temperatures and pressures must be enormous, while some stars which varied their light (variable stars) were found to be alternately expanding and contracting, and doing so rapidly. This did not seem to fit in with a theory of energy from contraction, and astronomers began to turn to the atoms and the tiny particles which Ernest Rutherford (1871-1937), Niels Bohr (1885-1962) and others had discovered, to see if these could produce the required energy in themselves.

35 Eddington on sub-atomic stellar energy

In 1927 Arthur Eddington (1882-1944) described this new source of stellar energy:

This store of energy is, with insignificant exception, energy of constitution of atoms and electrons; that is to say, subatomic energy. Most of it is inherent in the constitution of the electrons and protons—the elementary negative and positive electric charges—out of which matter is built; so that it cannot be set free unless these are destroyed. The main store of energy in a star cannot be used for radiation unless the matter composing the star is being annihilated.

It is possible that the star may have a long enough life without raiding the main energy store. A small part of the store can be released by a process less drastic than annihilation of matter, and this might be sufficient to keep the sun burning for 10,000,000,000 years or so, which is perhaps as long as we can reasonably require. The less drastic process is transmutation of the elements. Thus we have reached a point where a choice lies open before us; we can either pin our faith to transmutation of the elements, contenting

ourselves with a rather cramped time-scale, or we can assume the annihilation of matter, which gives a very ample time-scale. But at present I can see no possibility of a third choice. Let me run over the argument again. First we found that energy of contraction was hopelessly inadequate; then we found that the energy must be released in the interior of the star, so so that it comes from an internal not external source; now we take stock of the whole internal store of energy. No supply of any importance is found until we come to consider the electrons and atomic nuclei; here a reasonable amount can be released by re-grouping the protons and electrons in the atomic nuclei (transmutation of elements), and a much greater amount by annihilating them.

Transmutation of the elements—so long the dream of the alchemist—is realized in the transformation of radio-active substances. Uranium turns slowly into, a mixture of lead and helium. But none of the known radio-active processes liberate anything like enough energy to maintain the sun's heat. The only important release of energy by transmutation occurs at the very beginning of the evolution of the elements.

We must start with hydrogen. The hydrogen atom consists simply of a positive and negative charge, a proton for the nucleus plus a planet electron. Let us call its mass 1. Four hydrogen atoms will make a helium atom. If the mass of the helium atom were exactly 4, that would show that all the energy of the hydrogen atoms remained in the helium atom. But actually the mass is 3.97; so that energy of mass 0.03 must have escaped during the formation of helium from hydrogen. By annihilating 4 grammes of hydrogen we should have released 4 grammes of

47 *An enlarged section of the nebula NGC 2237 in Monoceros, showing some dark spots that are thought to be nodules of gaseous material condensing to form stars. Photographed with the 48in Schmidt telescope at Palomar*

66

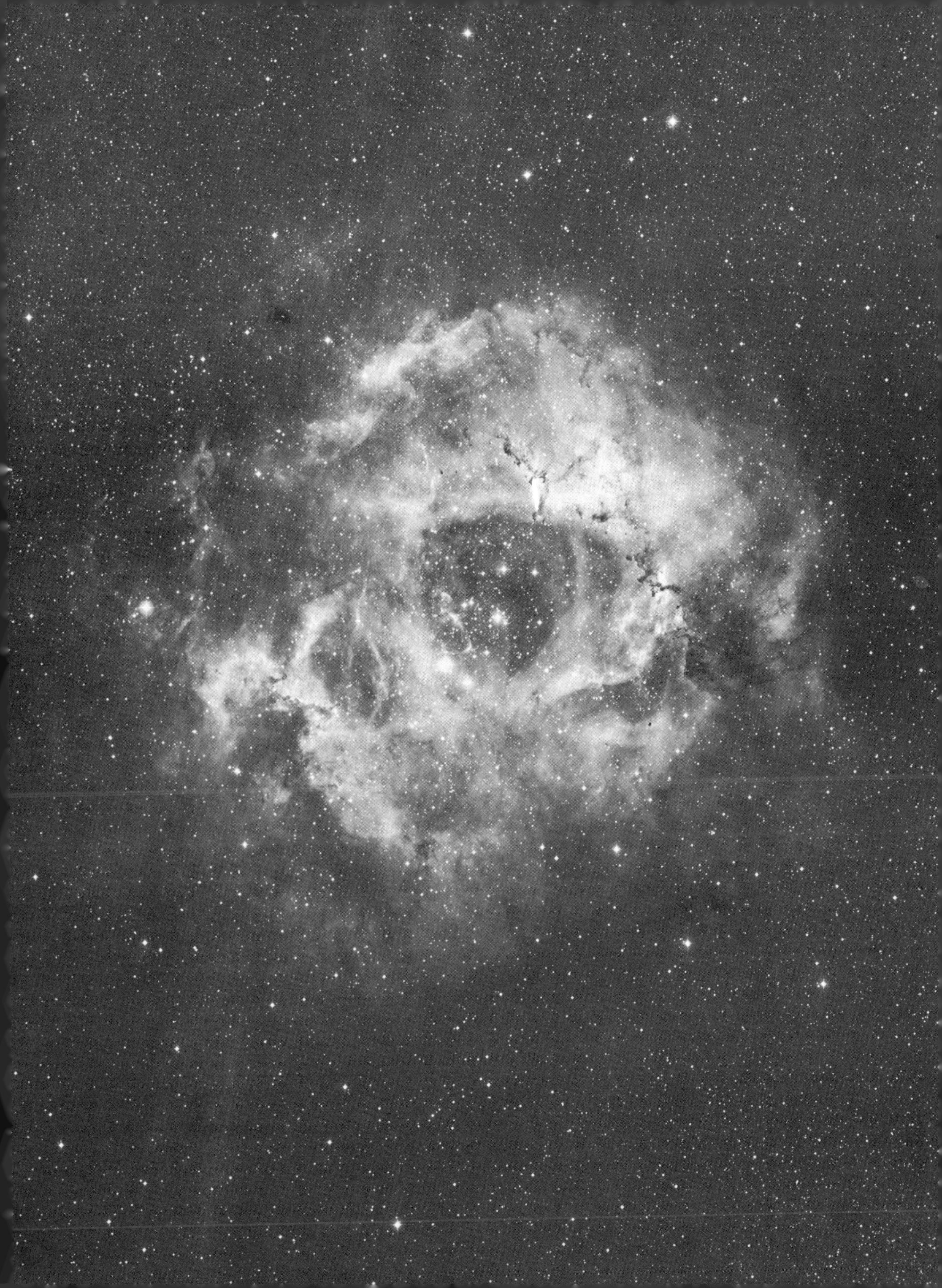

energy, but by transmuting it into helium we release 0.03 grammes of energy. Either process might be used to furnish the sun's heat though, as we have already stated, the second gives a much smaller supply.

The release of energy occurs because in the helium atom only two of the four electrons remain as planet electrons, the other two being cemented with the four protons close together in the helium nucleus. In bringing positive and negative charges close together you cause a change of the energy of the electric field, and release electrical energy which spreads away as ether-waves. That is where the 0.03 grammes of energy has gone. The star can absorb these ether-waves and utilize them as heat.

Eddington was here discussing the energy that could theoretically be produced by the synthesis of lighter atoms into heavier ones, now familiar in the hydrogen bomb and known as thermonuclear reactions. The transformation of matter into energy was a consequence of Einstein's theory of relativity, and once Eddington had given the clue, the precise mechanism required working out. To do this involved theoretical nuclear physics and no little computation.

36 Bethe on energy production in stars

In 1932 Robert d'E. Atkinson determined some details and in 1938 Hans Bethe was able to give a full description of the basic processes of stellar energy generation. To appreciate something of his achievement it may be as well to remind ourselves of the fundamental facts at Bethe's disposal. First, hydrogen atoms (H), the simplest of all atoms, composed of a nucleus with one orbiting electron, the nucleus usually being itself a proton (with a positive electric charge to balance the negative charge of the electron). Second, neutrons, having no electric charge and equal in mass to a proton. Third, isotopes—atoms with the same numbers of electrons and protons (and hence chemically the same) but with different numbers of neutrons, so that the atoms have different weights. Thus there are isotopes of carbon like C^{12}—carbon atoms whose nuclei have 6 protons and 6 neutrons—and C^{13}, with 6 protons and 7 neutrons, isotopes of nitrogen (N) and oxygen (O), and even of hydrogen where instead of a single proton there is a neutron as well; this is known as heavy hydrogen, and its nuclei as deuterons. Fourth, positrons (ε^+)—small light particles that can perhaps best be described as electrons with a positive electric charge—and, fifth, alpha (α) particles that are the nuclei of helium (He) atoms which, in their ordinary state, are orbited by two electrons:

It is shown that the *most important source of energy in ordinary stars is the reactions of carbon and nitrogen with protons.* These reactions form a cycle in which the original nucleus is reproduced, *viz* $C^{12}+H=N^{13}$, $N^{13}=C^{13}+\varepsilon^+$, $C^{13}+H=N^{14}$, $N^{14}+H=O^{15}$, $O^{15}=N^{15}+\varepsilon^+$, $N^{15}+H=C^{12}+He^4$. Thus carbon and nitrogen merely serve as catalysts for the combination of four protons (and two electrons) into an α-particle.

The carbon-nitrogen reactions are unique in their cyclical character. For all nuclei lighter than carbon, reaction with protons will lead to the emission of an α-particle so that the original nucleus is permanently destroyed. For all nuclei heavier than fluorine, only radiative capture of the protons occurs, also destroying the original nucleus. Oxygen and fluorine reactions mostly lead back to nitrogen. Besides, these heavier nuclei react much more slowly than C and N and are therefore unimportant for the energy production.

The agreement of the carbon-nitrogen reactions with observational data is excellent. In order to give the correct energy evolution in the sun, the central temperature of the sun would have to be 18.5 million degrees while integration of the Eddington equations gives 19. For the brilliant star Y Cygni the corresponding figures are 30 and 32. This good agreement holds for all bright stars of the main sequence, but, of course, not for giants.

For fainter stars, with lower central temperatures,

the reaction $H + H = D + \varepsilon^+$ and the reactions following it, are believed to be mainly responsible for the energy production.

It is shown further that *no elements heavier than He^4 can be built up in ordinary stars*. This is due to the fact, mentioned above, that all elements up to boron are disintegrated by proton bombardment (α-emission!) rather than built up (by radiative capture). The instability of Be^8 reduces the formation of heavier elements still further. The production of neutrons in stars is likewise negligible. The heavier elements found in stars must therefore have existed already when the star was formed.

With this important paper, giving reasons for the relationship astronomers had found between the massiveness of a star and its luminosity, the astronomer had moved into a new world where the behaviour of the stars became part of nuclear physics, and was far removed from Aristotle's universe where a celestial essence held sway.

CHAPTER FOUR

The Nebulae

In the sky, without a telescope, one can observe an apparent multitude of stars, and here and there one or two hazy patches. When drawing up his catalogue of stars, Hipparchos noted two objects of this kind, while Ptolemy observed five more, and gave all seven in the catalogue in his Almagest. *Today we recognise only one of these as a truly hazy 'nebula'; for it is only with the telescope that any real study of such objects has been possible.*

37 Galileo on nebulae

In 1609 when Galileo turned his telescope to the skies, he observed various nebulae, and saw some as clusters of stars. He came to an important conclusion:

The next object which I have observed is the essence or substance of the Milky Way. By the aid of a telescope any one may behold this in a manner which so distinctly appeals to the senses that all the disputes which have tormented philosophers through so many ages are exploded at once by the irrefragable evidence of our eyes, and we are freed from wordy disputes upon this subject, for the Galaxy is nothing else but a mass of innumerable stars planted together in clusters. Upon whatever part of it you direct the telescope straightway a vast crowd of stars presents itself to view; many of them are tolerably large and extremely bright, but the number of small ones is quite beyond determination.

And whereas that milky brightness, like the brightness of a white cloud, is not only to be seen in the Milky Way, but several spots of a similar colour shine faintly here and there in the heavens, if you turn the telescope upon any of them you will find a cluster of stars packed close together. Further—and you will be more surprised at this—the stars which have been called by every one of the astronomers up to this day *nebulous*, are groups of small stars set thick together in a wonderful way, and although each one of them on account of its smallness, or its immense distance from us, escapes our sight, from the com-

mingling of their rays there arises that brightness which has hitherto been believed to be the denser part of the heavens, able to reflect the rays of the stars or the Sun.

38 Halley on nebulae

However, just over a century later, using improved instruments, Halley proposed an entirely different interpretation:

Not less wonderful are certain luminous Spots or Patches, which discover themselves only by the Telescope, and appear to the naked Eye like small fixt Stars; but in reality are nothing else but the Light coming from an extraordinary great Space in the Ether; through which a lucid *Medium* is diffused, that shines with its own proper Lustre. This seems fully to reconcile that Difficulty *Moses* gives of the

48 *Observing with a wooden tubed refracting telescope of 30ft focal length at Hevelius' observatory in Danzig, c 1640. Mounted on a post with block and tackle, the instrument was not convenient to use. Because of optical deficiencies, long focal lengths had to be adopted, and Hevelius even went so far as to construct a large refractor with a focal length of more than 150ft. It was cumbersome and suffered some bending of its long single-sided 'tube'; telescopes like that illustrated here were more usually adopted. From Hevelius,* Selenographia *(Danzig 1647)*

Creation, alledging that Light could not be created without the Sun. But in the following Instances the contrary is manifest; for some of these bright Spots discover no sign of a Star in the middle of them; and the irregular Form of those that have shews them not to proceed from the Illumination of a Central Body. These are, as the aforesaid New Stars, six in Number, all which we will describe in the order of Time, as they were discovered; giving their Places in the Sphere of Fix't Stars, to enable the Curious, who are furnished with good Telescopes, to take the Satisfaction of contemplating them.

39 Messier's catalogue of nebulae and star clusters

But not everyone accepted Halley's opinion, and indeed 20 years later William Derham (1657-1735) in a paper in the Philosophical Transactions, *still suggested that the nebulae were holes in the sky through which one could glimpse a 'Region of Light, beyond the Fix't Stars', a view that was old-fashioned even in the eighteenth century. Nevertheless, whatever they might think them to be, astronomers continued to list them, and the most notable catalogue was made in 1781 by Charles Messier (1730-1817). Messier's great interest lay in observing comets which, until close to the Sun, can easily be mistaken in a telescope for a dim nebula, and he drew up his catalogue to avoid one obvious source of confusion. That Messier is now remembered for his catalogue of nebulae—his numbering of the objects is still used in astronomy—and not for his cometary observations is one of the ironies of astronomical history:*

DATE   NO. DIAM.   DETAIL OF NEBULAE AND CLUSTERS

..   ..

1764

July 12 27   0° 4′   Nebula without star, discovered in Vulpecula, between the two forepaws and very near the star 14 of that constellation, 5th magnitude following Flamsteed;

DATE   NO. DIAM.   DETAIL OF NEBULAE AND CLUSTERS

it can be well seen in an ordinary telescope of three and a half feet: it appears oval shaped and contains no star. M. Messier has reported its position on the Chart of the Comet of 1779 which will be engraved for the volume of the Acad. of the same year. Reviewed 1781 January 31.

..   ..

1764

Aug. 3   31   0° 40′   The fine nebula in Andromeda's girdle, shaped like a spindle; M. Messier has examined it with three different instruments and has not found a single star there: it resembles two cones or pyramids of light, base-opposed, of which the axis is in the direction north-west to south-east; the two points of light or the two apexes are separated by 40 minutes of arc; the common base of the pyramids, 15 minutes. This nebula was discovered in 1612 by Simon Marius and observed afterwards by different Astronomers. M. le Gentil has given a drawing of it in the Memoirs of the Academy of 1759, page 453. It is reported in the English Atlas.

..   ..

But the greatest cataloguer of nebulae was William Herschel who, between 1786 and 1802, compiled three lists containing more than 2,500 entries, and included star clusters as well as nebulae. His son John extended this work, adding 500 more items, and then observing in the southern hemisphere and adding over 1,700

*Catalogues (*IC*) were issued, and by then the total number of such objects had reached more than* 13,000. *Obviously then, there are a vast number of nebulae and*

49 *Telescopes at the Radcliffe Observatory, Oxford, c 1814. On the far left is one of William Herschel's reflectors of 10ft focal length and with a mirror of some 8in diameter. In the centre is a refractor of some 2in aperture with a wooden tube and, on the right, an equatorially mounted refractor with a brass tube and an improved colour-corrected (achromatic) object-glass of two components. From an aquatint published in 1814 in London by Rudolph Ackermann (1764-1834) in his* History of Oxford

additional objects. In 1864 *the complete list comprising some* 5,000 *entries was published. But this did not end the search, and in* 1888 *John Dreyer* (1852-1926) *brought the list up to date and called it the* New General Catalogue (*NGC*): *in* 1895 *and* 1908 *two* Index

50 *William Herschel's giant reflector with a tube 40ft in length. In this instrument Herschel adopted his own 'front view' method, in which the observer gazed via an eyepiece straight down the tube, the main mirror being slightly tilted to allow for this. The large diameter of the main mirror—48in—enabled this method to be practical (in a much smaller telescope, the observer's head or a portion of it would obstruct too much light from entering the instrument). Herschel's front view obviated reflection from a secondary mirror, as in Newton's design, which Herschel had usually adopted, and so avoided an additional source of loss of light. This copperplate engraving shows one of the two circular brick piers on which the telescope rotated. From the* Encyclopaedia Londinensis (*London 1820*)

*clusters, all of which are usually referred to by their
catalogue numbers—thus M 27, the one in Vulpecula, is
also known as NGC 6853, M 31 in Andromeda as
NGC 224, and so on. But what are they?*

40 William Herschel on the real existence of nebulae

*Astronomers soon agreed that Derham was mistaken,
but was Halley correct, or was Galileo nearer the truth?*

*William Herschel was clearly much involved with the
question and he attempted to see whether, with more
powerful telescopes, the hazy patches would resolve into
myriads of separate stars. Herschel had an astounding
ability for making telescopes: he constructed reflectors,
his largest having a mirror of 4ft diameter and a tube
40ft long (figure 50). Yet even with this gargantuan
instrument the question could not be answered:*

A KNOWLEDGE of the construction of the heavens
has always been the ultimate object of my observations, and having been many years engaged in
applying my forty, twenty and large ten feet
telescopes, on account of their great space-penetrating
power to review the most interesting objects discovered in my sweeps, as well as those which had
before been communicated to the public in the
Connoissance des Temps, for 1784, I find that by

arranging these objects in a certain successive regular order, they may be viewed in a new light, and, if I am not mistaken, an examination of them will lead to consequences which cannot be indifferent to an inquiring mind.

Impressed with an idea that nebulæ properly speaking were clusters of stars, I used to call the nebulosity of which some were composed, when it was of a certain appearance, *resolvable*; but when I perceived that additional light, so far from resolving these nebulæ into stars, seemed to prove that their nebulosity was not different from what I had called milky, this conception was set aside as erroneous. In consequence of this, such nebulæ as afterwards were suspected to consist of stars, or in which a few might be seen, were called easily *resolvable*; but even this expression must be received with caution, because an object may not only contain stars, but also nebulosity not composed of them.

41 Rosse on the nature of the nebulae

In 1845 William Parsons, third Earl of Rosse (1800-67), completed a telescope even larger than Herschel's, and set it to work at his country seat Birr Castle, Parsonstown, in Offaly in Ireland. Known as the 'Leviathan of Parsonstown', it boasted a mirror of 6ft diameter, and was certainly an astounding instrument for the time (figure 51). Although some nebulae were still unresolved even with this instrument, Rosse remained reasonably sure that what his giant telescope could not achieve, even larger instruments might be able to resolve.

The letter 'r' has been occasionally added to the description [of nebulae], and always in the same sense as that in which HERSCHEL employed it: I do not, however, attach much importance to the expression of opinion it conveys, because the question of resolvability can only be successfully investigated when the air is steady and the speculum [main mirror] in fine order. In the early observations with the 6-feet telescope we had the advantage of a very fine

52 *The Crossley reflector. Constructed by Andrew Ainslie Common (1841-1903), the telescope had a silver-on-glass mirror of 36in diameter, and also made use of late nineteenth-century engineering techniques to provide an efficient mounting. In 1885 Common sold this telescope to Edward Crossley who, in turn, presented it to Lick Observatory in California, where with the mounting slightly modified, it was used by James Keeler (1857-1900). His success with the instrument placed the reflector in the forefront of large telescopes and led to the construction, in the twentieth century, of reflectors of very large aperture*

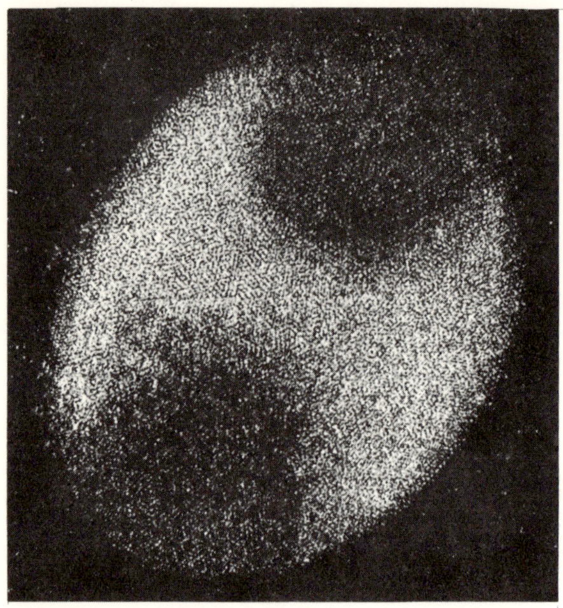

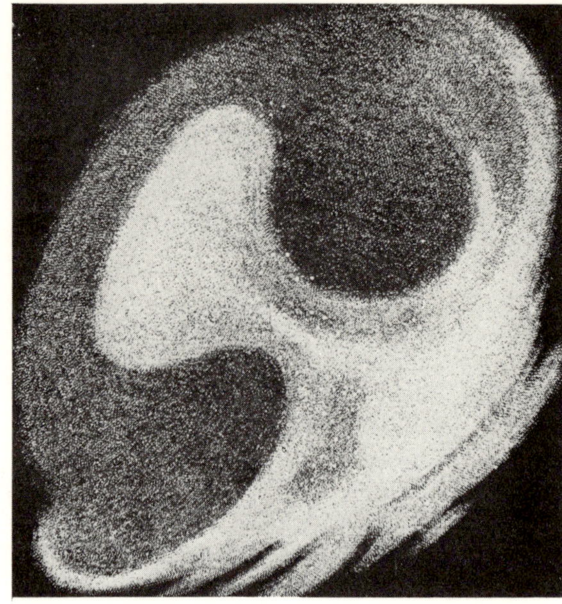

53 *The 'Dumb-bell' nebula (M27) in the constellation Vulpecula, as observed by John Herschel. From an engraving in Amédée Guillemin,* The Heavens *(London 1867)*

54 *The 'Dumb-bell' nebula (M27) as Rosse recorded it. From an engraving in Amédée Guillemin,* The Heavens *(London 1867). Compare with figure 53*

speculum; it had been polished at the close of a long series of experiments with 3-feet specula, when by practice every refinement of manipulation was fresh in the recollection; there were also at that time several very good nights; and many nebulæ were resolved. Very soon after, the spiral form of arrangement was detected; and our attention was then directed to the form of nebulæ, the question of resolvability being a secondary object. In the mean time the speculum, which had been frequently dewed and occasionally cleaned, had lost its fine edge, and was no longer in a state to deal with the question of resolvability. Our aim was to trace out faint details, and in that respect also the speculum had lost much of its power. It was therefore repolished, and, though less perfect than before, did the work we required well. Since that, we have had perhaps two or three specula as perfect as the first one; but the mass of observations have been made with specula considerably inferior to it, and, I am sorry to say, very often not as bright as they should have been. The removing a 6-feet speculum from the telescope to the laboratory, repolishing it, perhaps several times, and replacing it, is a serious operation, and has often been too long postponed. While the telescope was in constant use in all weathers, it would have been a hopeless task to attempt to keep it in a state fit for the resolution of nebulae, and the attempt was not made. I may, perhaps, mention that with the 3-feet speculum in fine order I have often detected resolvability when there was no trace of it with the 6-feet speculum in its ordinary working-condition.

The question of resolvability, therefore, I think, must remain to be taken up separately, when the

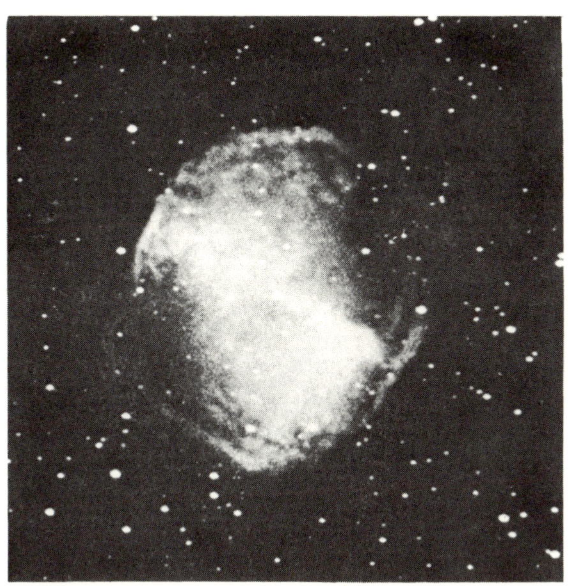

55 *The 'Dumb-bell' nebula (M27), photographed by James Keeler with the Crossley reflector (figure 52). From* Publications of the Lick Observatory, *volume 8 (1908)*

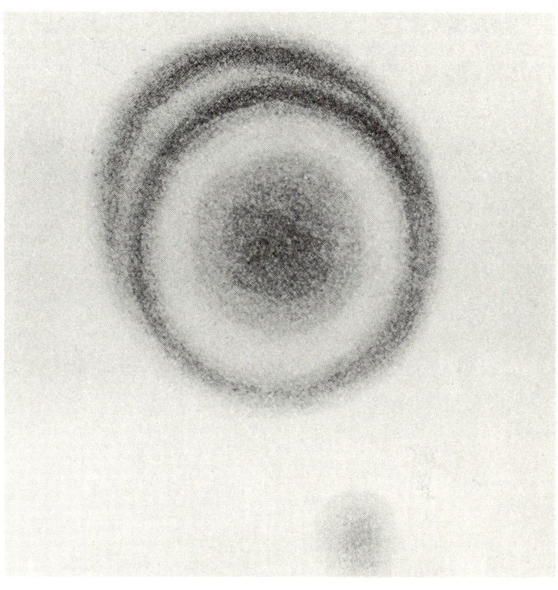

56 *The galaxy in the constellation Canes Venatici (M51) as recorded by John Herschel. From François Arago,* Astronomie Populaire, *volume 1, (Paris 1857)*

finest instrumental means are available, and when it may no longer be necessary to subject the specula to the wear and tear of constant work.

42 Huggins on the spectra of nebulae

The spiral form that Rosse's large telescope showed up (figure 57) was a matter of considerable curiosity, but it seemed at the time only to be a new type of nebula, and nebulae came in an immense variety of shapes—only many years later was the significance of the spiral appreciated. But the problem of resolution of nebulae into stars was not solved, even by the 'Leviathan'; yet a solution did come, and from an unexpected quarter— the observations of Sir William Huggins:

On the evening of the 29th August I directed the telescope for the first time to a planetary nebula in Draco.

I looked into the spectroscope. No spectrum such as I expected! A single bright line only! At first I suspected some displacement of the prism, and that I was looking at a reflection of the illuminated slit from one of its faces. This thought was scarcely more than momentary; then the true interpretation flashed upon me. The light from the nebula was monochromatic; and so, unlike any other light I had as yet subjected to prismatic examination, could not be extended out to form a complete spectrum. After passing through the two prisms it remained concentrated into a single bright line, having a width corresponding to the width of the slit, and occupying in the instrument a position at that part of the spectrum to which its light belongs in refrangibility. A little closer looking showed two other bright lines on the side towards the blue, all the three lines being separated by intervals relatively dark.

The riddle of the nebulae was solved. The answer, which had come to us in the light itself, read: Not an aggregation of stars, but a luminous gas.

The planetary nebula Huggins had observed was called 'planetary' because in a telescope it appeared as a small disk just like a planet, but it had nothing to do with planets and was in fact a spherical shell of gas. By 1866 Huggins had examined the spectra of 60 nebulae and had found that one-third were gaseous, but two-thirds were not. We now know that these other objects are great concourses of stars, and this was the reason the Herschels and Rosse, as well as others, had such difficulty in interpreting their observations.

43 Huggins on the red-shift of spectral lines

But having solved one problem, Huggins observed some additional evidence which was to prove of singular importance when the 'starry' nebulae were studied in greater detail in the twentieth century. This was an examination of the position of the spectral lines of stars. In 1842 the Austrian physicist Christian Doppler (1803-53) had attempted to explain colour changes that had been observed in variable stars by assuming that they were moving towards or away from the Earth, since if their velocities were large then this motion would affect the frequency with which their light waves impinged on the observer: any such effect would then cause the apparent colour of the light to be different. If the stars moved away, their light would become redder, if they moved towards the Earth, the light would become bluer.

But Doppler was wrong, because a star radiates not only light but also rays that extend beyond the ends of the visible spectrum. Beyond the violet end lie ultra-violet rays and then still shorter waves—X-rays and γ- (gamma) rays. At the long end of the spectrum, below the red, lie the infra-red rays, that were first recognised by William Herschel (figure 62). If a star is moving towards us, its light would not be made bluer, as Doppler supposed, because some of the infra-red would be shifted into the visible range, becoming red, and so compensate for the shift of the visible light away from the red. However, Hippolyte Fizeau (1819-96) pointed

57 *The galaxy in Canes Venatici (M51) as recorded by Rosse, and showing its spiral nature. From* Phil Trans, *volume 160 (1850)*

58 *The galaxy in Canes Venatici (M51) photographed by James Keeler with the Crossley reflector (figure 52). From* Publications of the Lick Observatory; *volume 8 (1908)*

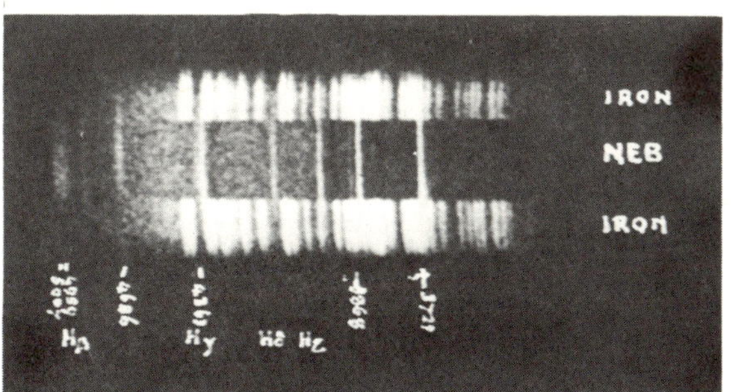

59 Above: *The spectrum of the 'Dumb-bell' nebula (M27) photographed by Max Wolf. At the top and bottom of the central spectrum are comparison spectra made before and after the nebula spectrum was exposed, using light from a spark generated between two iron rods, as this gives a series of lines along the entire spectrum. The bright lines against a dark background show that the nebula is a gas Below: The spectrum of a galaxy photographed by Max Wolf. This shows dark lines against a bright background, and thus indicates the presence of stars. As in the photograph above, an iron spark spectrum is exposed on top of and underneath the spectrum of the galaxy*

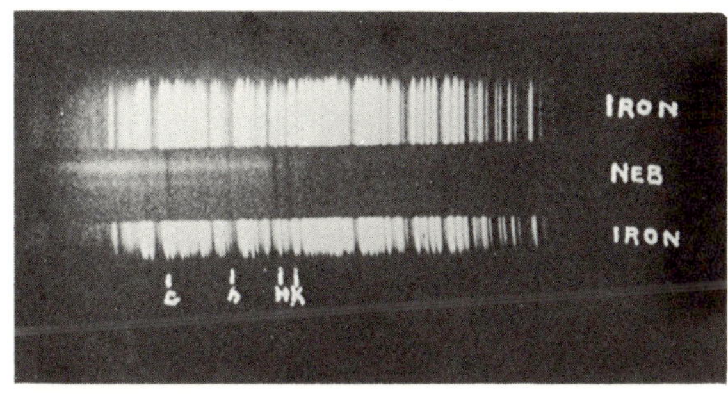

out that while there would be no colour change, line-of-sight motion should give a shift of the spectral lines towards the red if the star is moving away, and towards the blue if it is moving towards the observer. This Doppler-Fizeau effect—rather unfairly known simply as the Doppler effect—is of considerable significance because stars only appear as points of light in even the largest telescope, and give no ready visual evidence of any line-of-sight movement:

60 *The Hale 200in diameter reflector on Palomar mountain, California, completed in 1948. Its mirror, made of Pyrex low-expansion glass, is coated with a highly reflecting film of aluminium. The tube of the instrument together with the cage for the observer weighs 140 tons, and the whole telescope represents a very high degree of optical and mechanical sophistication*

Sirius.—The brilliant light of this star and the great intensity of the four strong lines of its spectrum, make it especially suitable for such an examination [Doppler-Fizeau shift]. The low altitude of this star in our latitude limits the period in which it can be successfully observed to about one hour on each side of the meridian.

I have confined myself to comparisons of the strong line in the position of F, with the corresponding line of the spectrum of hydrogen. My first trials were made with hydrogen at the ordinary atmospheric pressure; the width of the band of hydrogen, under these circumstances, was greater than the line of Sirius. This line in Sirius, from some cause, is narrower relatively to the length of the spectrum, when considerable dispersion and a narrow slit are employed, than when the image of the star, rendered linear by a cylindrical lens, is observed with a single prism.

· · ·

As it was obviously impossible to determine with the required accuracy the coincidence of the line of Sirius when the much broader band of hydrogen at the ordinary pressure was compared with it, I em-

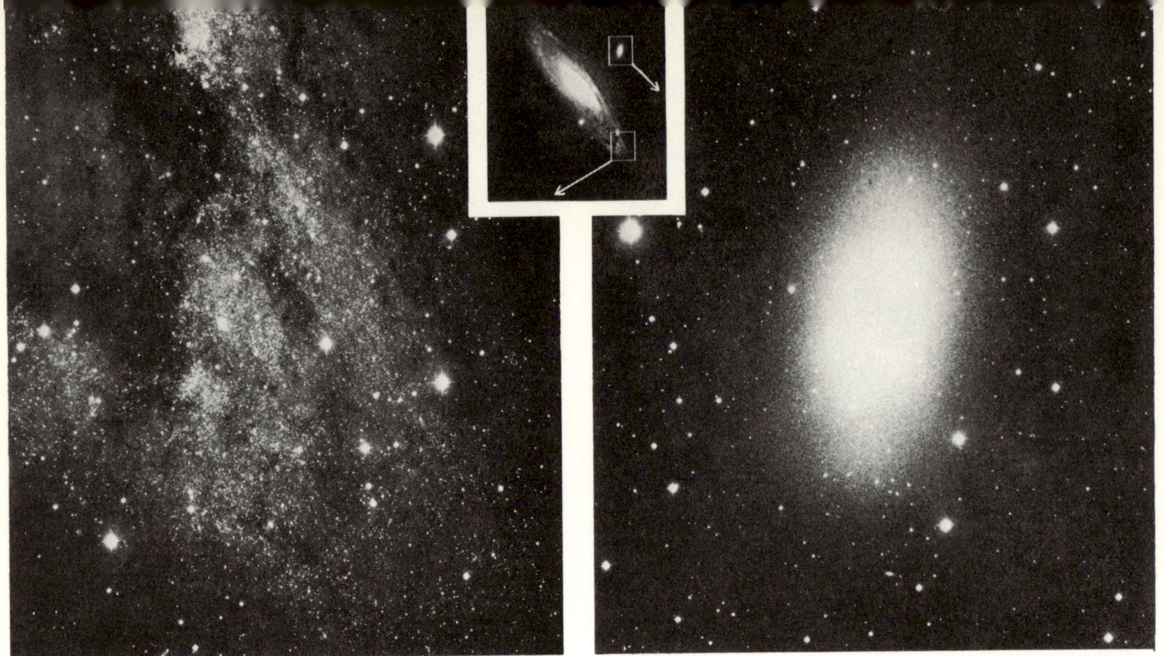

61 *The spiral galaxy in Andromeda (M31), with its companion galaxy—an elliptical (NGC 205)—photographed by the 200in telescope. Both are shown in the central photograph, while on the left, giant and supergiant blue stars in one of the spiral arms of M31 may be seen resolved into separate points in a photograph taken in blue light. On the right, in a photograph taken in yellow light, separate red coloured stars can be detected in NGC 205. Walter Baade (1893-1960) classified these two classes of stars, which were drawn to his attention by this photograph. The blue ones, found in the spiral arm of M31 and of other spiral galaxies, he called stars of Population I, and they are thought to be younger than the red stars, found in the elliptical galaxy NGC 205 and in other ellipticals, as well as in the nucleus of a spiral galaxy like M31 and of our own Galaxy, which he called Population II. The very bright stars on the photographs are stars of our own Galaxy*

ployed a vacuum-tube fixed before the object-glass. In all these observations the slit used was as narrow as possible. The air at the time of the present observations was more favourable than usual, and the line in Sirius was seen with great distinctness. The line from the spark appeared, in comparison, very narrow, not more than about one-fifth of the width of the line of Sirius. . . . I was unable to measure directly the distance between the centre of the line of hydrogen and that of the line in the spectrum of Sirius, but several very careful estimations by means of the micrometer give a value for that distance of 0.040 of the micrometer-head. This value is probably not in error by so much as its eighth part [figure 63].

. . . . . .

From these observations it may, I think, be concluded that the substance in Sirius which produces the strong lines is really hydrogen, as was stated by Dr. MILLER and myself in our former paper. Further, that the aggregate result of the motions of the star and the earth in space, at the time when the observations were made, was to degrade the refrangibility [another way of referring to the frequency of light] of the line in Sirius by an amount corresponding to 0.040 of the micrometer-screw . . . [and] the observed alteration in period of the line in Sirius will indicate a motion of recession existing between the earth and the star of 41.4 miles per second. . . .

There remains unaccounted for a motion of recession from the Earth amounting to 29.4 miles per second, which we appear to be entitled to attribute to Sirius.

Thus, the stage was now set to look again at the layout of the universe, the kinds of objects in it, and their motions both across the sky and in the line-of-sight.

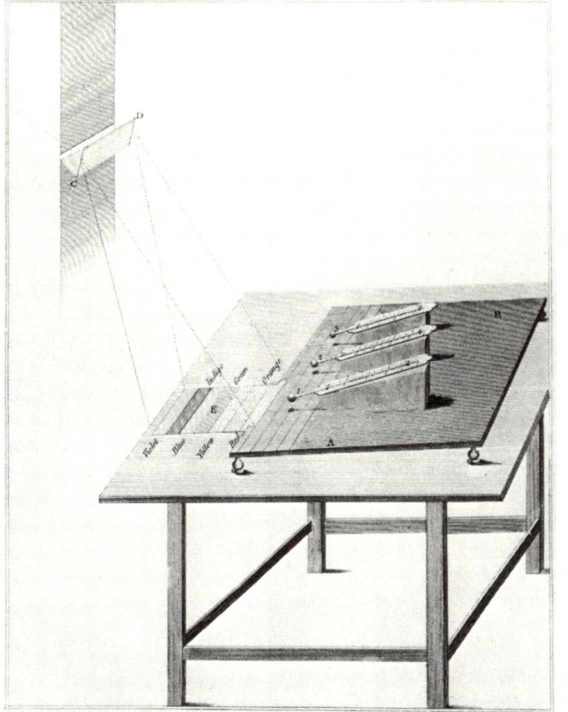

62 *William Herschel's detection of infra-red radiation, using a series of thermometers. From his paper 'Experiments on the Refrangibility of the invisible Rays of the Sun', in* Phil Trans *volume 90 (1800)*

63 *The red shift of a spectral line of Sirius, detected by William Huggins in 1868.* Phil Trans, *volume 158 (1868) (see extract 43)*

The Layout of the Universe

What is the universe really like? How far does it extend? Is it finite or infinite? Questions like these are perennial, but the answers we give depend not only on the facts as they are known, but also on the outlook of the age, on the climate of thought at a particular time. To the Greeks the facts of observations, which were few and simple, had to be married to a grand philosophical scheme embracing the whole of creation, and possessing certain aesthetic qualities that were conceived of as an integral part of it:

44 Aristotle on the shape of the universe

The shape of the heaven is of necessity of spherical; for that is the shape most appropriate to its substance and also by nature primary.

First, let us consider generally which shape is primary among planes and solids alike. Every plane figure must be either rectilinear or curvilinear. Now the rectilinear is bounded by more than one line, the curvilinear by one only. But since in any kind the one is naturally prior to the many and the simple to the complex, the circle will be the first of the plane figures. Again, if by complete, as previously defined [complete in itself: self-contained] we mean a thing outside which no part of itself can be found, and if addition is always possible to the straight line but never to the circular, clearly the line which embraces the circle is complete. If then the complete is prior to the incomplete, it follows on this ground also that the circle is the primary among figures. And the sphere holds the same position among solids. For it alone is embraced by a single surface, while rectilinear solids have several. The sphere is among solids what the circle is among plane figures. . . .

Again, since the whole revolves, palpably and by assumption, in a circle, and since it has been shown that outside the farthest circumference there is neither void nor place, from these grounds also it will follow necessarily that the heaven is spherical. For if it is to be rectilinear in shape, it will follow that there is

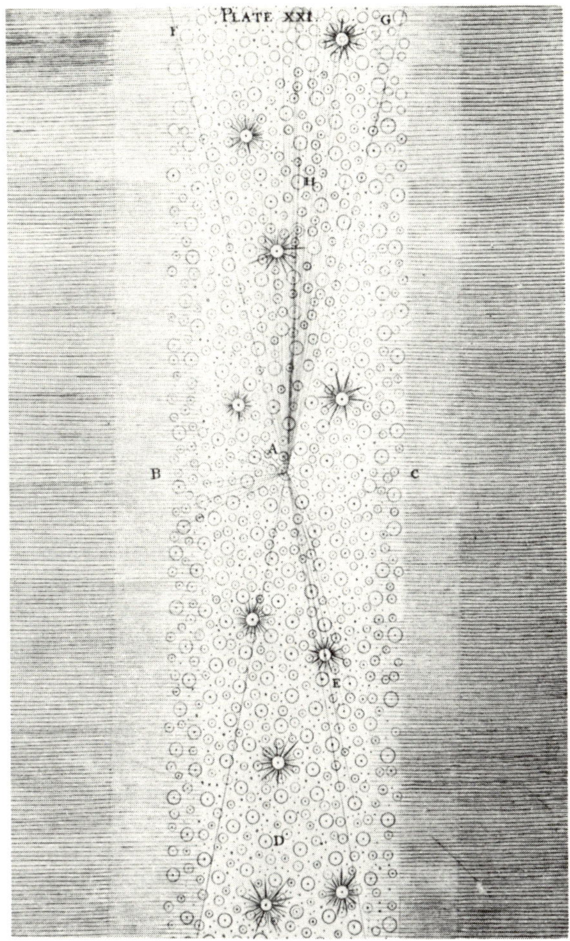

64 *Thomas Wright's conception of the layout of the* *universe, from his* An Original Theory or New Hypothesis of the Universe *(London 1750) (see extract 45)*

place or body and void without it. For a rectilinear figure as it resolves never continues in the same room, but where formerly was body, is now none, and where now is none, body will be in a moment because of the projection of the corners. Similarly, if the world had some other figure with unequal radii, if, for

instance, it were lentiform, or oviform, in every case we should have to admit space and void outside the moving body, because the whole body would not always occupy the same room.

45 Wright on the layout of the universe

Eighteen-hundred years later, the Copernican system, with the Sun at the centre of the universe, still retained the spherical Aristotelian universe, and only in England was it suggested that the cosmos was different, Thomas Digges turning to an infinite universe. Halley considered the question, and also concluded that the universe was infinite, arguing that otherwise it would have a centre to which all matter would gravitate. He also assumed that, on the average, the stars were evenly distributed in space. However, in 1750 Thomas Wright (1711-86) published AN ORIGINAL THEORY OR NEW HYPOTHESIS OF THE UNIVERSE:

Let us imagine a vast infinite Gulph, or Medium, every Way extended like a plane, and inclosed between two Surfaces, nearly even on both Sides, but of such a Depth as to occupy a Space equal to the double Radius, or Diameter of the visible Creation, that is to take in one of the smallest Stars each Way, from the middle Station, perpendicular to the Plane's Direction, and, as near as possible, according to our Idea of their true Distance.

. . . Now in this Space let us imagine all the Stars scattered promiscuously, but at such an adjusted Distance from one another, as to fill up the whole Medium with a kind of regular Irregularity of Objects. And next let us consider what the Consequence would be to an Eye situated near the Center Point, or any where about the middle Plane, as at Point A.

46 William Herschel on the 'Construction of the Heavens'

William Herschel also tried to fathom the general design of the universe, using his observations of numbers

It is very probable, that the great stratum, called the milky way, is that in which the sun is placed, though perhaps not in the very center of its thickness. We gather this from the appearance of the Galaxy, which seems to encompass the whole heavens, as it certainly must do if the sun is within the same. For, suppose a number of stars arranged between two parallel planes, indefinitely extended every way, but at a given considerable distance from each other; and, calling this a sidereal stratum, an eye placed somewhere within it will see all the stars in the

direction of the planes of the stratum projected into a great circle, which will appear lucid on account of the accumulation of the stars; while the rest of the heavens, at the sides, will only seem to be scattered over with constellations, more or less crowded according to the distance of the planes or number of stars contained in the thickness or sides of the stratum.

Thus in . . . [figure 65] an eye at S within the stratum *ab*, will see the stars in the direction of its length *ab*, or height *cd*, with all those in the intermediate situations projected into the lucid circle ACBD; while those in the sides *mv*, *nw*, will be seen scattered over the remaining part of the heavens at MVNW.

65 *William Herschel's idea of the distribution of the stars in space, from his paper in the* Phil Trans, *volume 74 (1784) (see extract 46)*

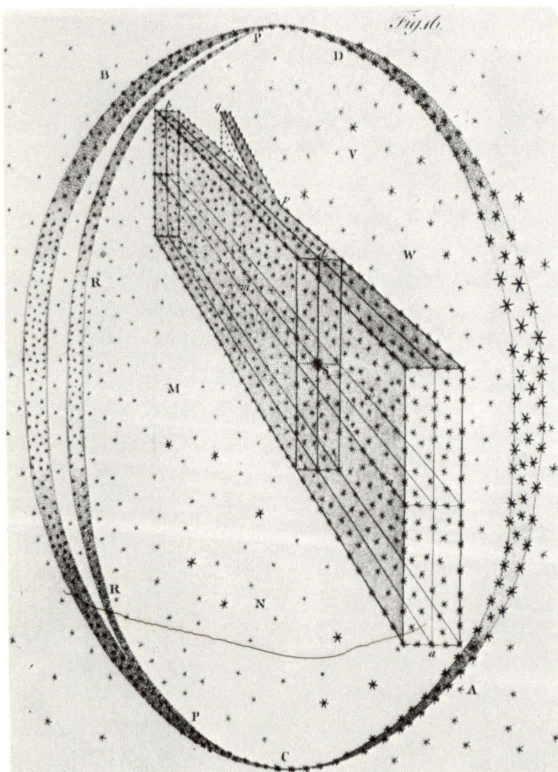

47 Kapteyn on two star streams

So far, the layout of the universe had assumed that the stars were stationary. Since Halley's day this had been known to be unlikely, and by the time William Herschel was proposing his scheme, their possible movement had to be taken into account, although it did not seem to affect the large-scale picture very markedly. However, once Huggins had determined the line-of-sight or 'radial' velocity of Sirius, the way was open to the determination of this velocity for other stars, and so by combining both radial and transverse motions (the motions Halley had observed), the real motions in space could be determined. A very careful investigation was made by Jacobus Kapteyn (1851–1922):

. . . This figure [figure 66] summarizes the more important points in regard to the question in hand. They show in compact form the results of a complete treatment of the proper motions of all the stars observed in both co-ordinates by Bradley (over 2,400 stars). These stars are distributed over two thirds of the whole of the sky. This surface has been divided up into twenty-eight areas.

From the stars contained on each area I have derived the distribution of the proper motions

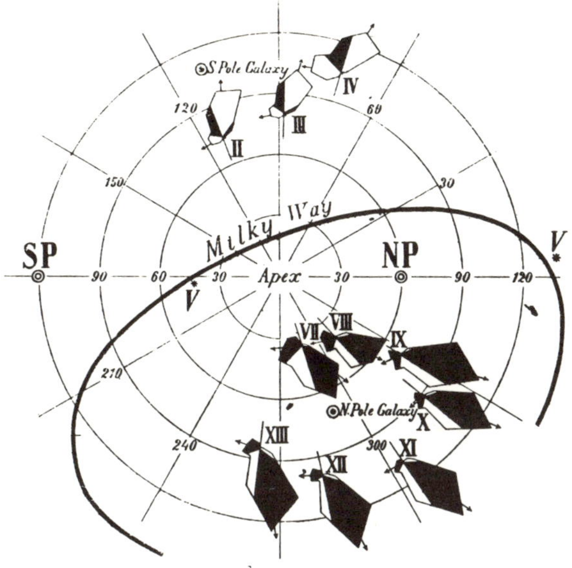

66 *One of Kapteyn's explanatory diagrams describing his two star-stream hypothesis. From his lecture given in 1905 to the annual meeting of the British Association for the Advancement of Science (see extract 47)*

corresponding to the center of the area. How this was done need not here be explained. The whole of the materials were thus embodied in twenty-eight figures, like those of . . . [figure 66], each of which shows at a glance the distribution of the proper motions for one particular region of the sky.

Not to overburden the plate, I have only included *ten* of the figures, for which the phenomenon to which I wish to draw your attention is most marked.

It is very suggestive that these lie all near to the poles of the Milky Way.

· · · · · · ·

When we see that the motions of a certain group of stars converge to a same point on the sphere we conclude either that the real motions of the stars are in reality parallel, or that the motion is only apparent and due to a motion of the observer in the opposite direction. As long as we have no fixed point of reference we cannot decide between the two.

When we see *two* groups of stars converging towards two *different* points the latter explanation fails, at least for one of the groups, because the observer can have but one motion.

From the facts set forth in what precedes we must, therefore, at once conclude that one of our sets must have a real systematic motion in respect to the other.

Kapteyn's discovery of two star streams led to a realisation that the whole star system was probably in motion, and the belief that the stars were all distributed to form a giant disk became increasingly accepted. But the positions and nature of the nebulae which Huggins had shown were composed of stars, were a problem. Were they near or far? Again, by the early years of the twentieth century many globular clusters of stars were known (figure 67). Where did these lie?

67 *A globular cluster (M3, NGC 5272) in Canes Venatici, as photographed in 1910 by George Willis Ritchey (1864-1945) in the United States*

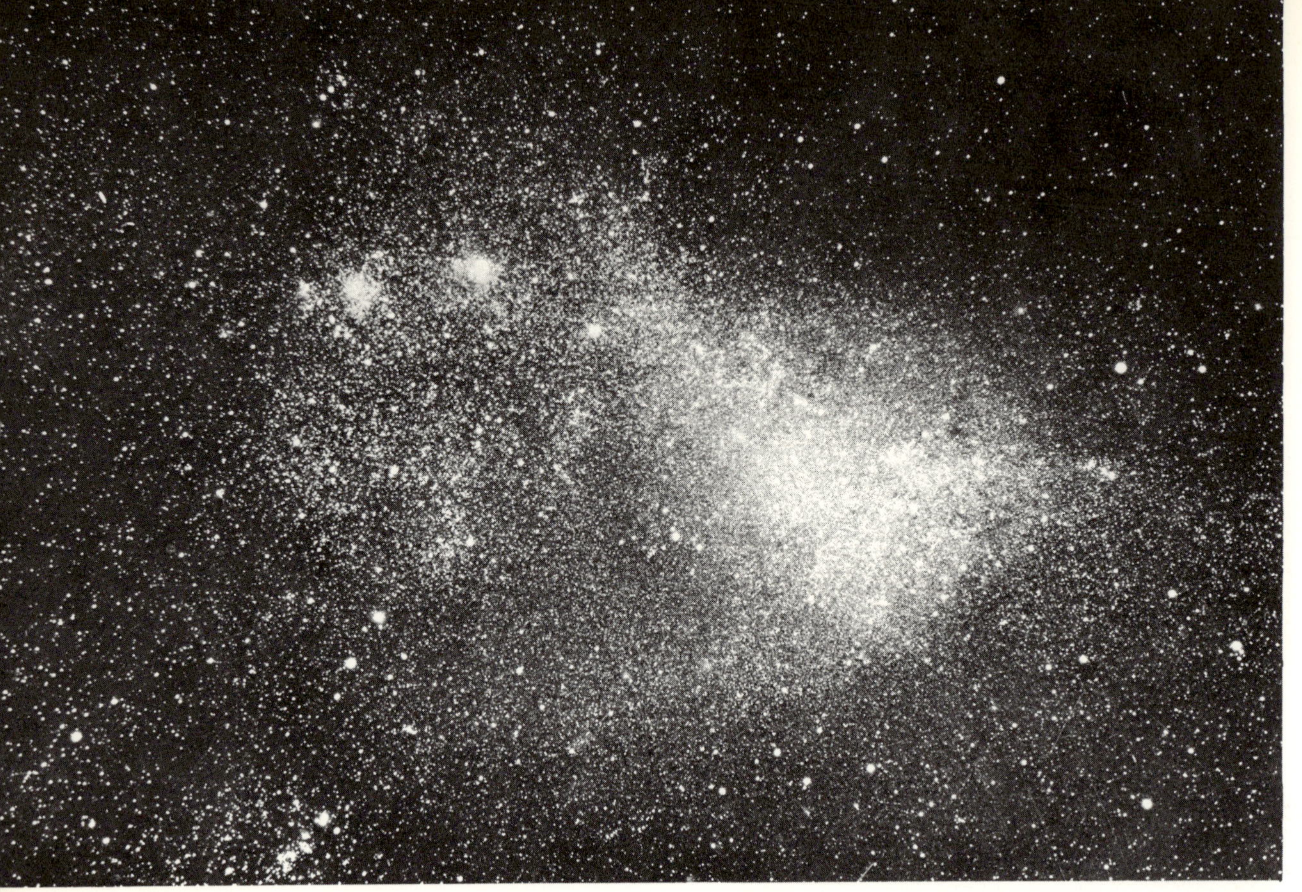

68 *The Small Magellanic Cloud, photographed in 1898 at the Harvard University observatory at Arequipa, Peru*

48 Henrietta Leavitt on variable stars in the smaller Magellanic cloud

In both cases the problem was to find some incontrovertible way of assessing great distances, and a new discovery was soon to provide this from observations made in the southern skies, where the Milky Way appears to possess two detached sections. Known as the Magellanic Clouds, the smaller (figure 68) was carefully examined by Henrietta Leavitt (1868-1921) and in 1912 she was able to announce a vitally important result:

Fifty-nine of the variables in the Small Magellanic Cloud were measured in 1904, using a provisional scale of magnitudes, and the periods of seventeen of them were published in *Harvard Annals* 60, No. 4, Table VI. They resemble the variables found in globular clusters, diminishing slowly in brightness, remaining near minimum for the greater part of the time, and increasing very rapidly to a brief maximum. . . . A remarkable relation between the brightness of these variables and the length of their periods will be noticed. In *Harvard Annals* 60, No. 4, attention was called to the fact that the brighter variables have the longer periods, but at that time it was felt that the number was too small to warrant the drawing of general conclusions. The periods of 8 additional variables which have been determined since that time, however, conform to the same law.

49 Shapley on the galaxy

The connection between brightness and length of period of variation—the period-luminosity relationship—proved a wonderful tool for measuring great distances. Harlow Shapley showed that the type of variable star Miss Leavitt had observed was typical of the short period kind of which δ (delta) Cephei was the most notable—a type that became known as 'Cepheid variables'. So if one observed a distant object—a globular cluster, say—

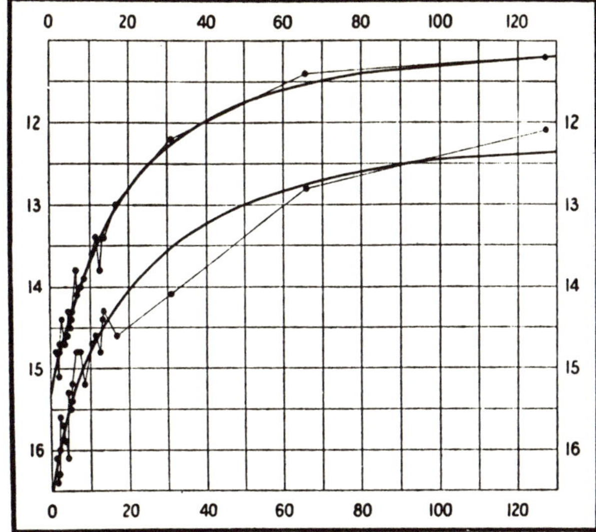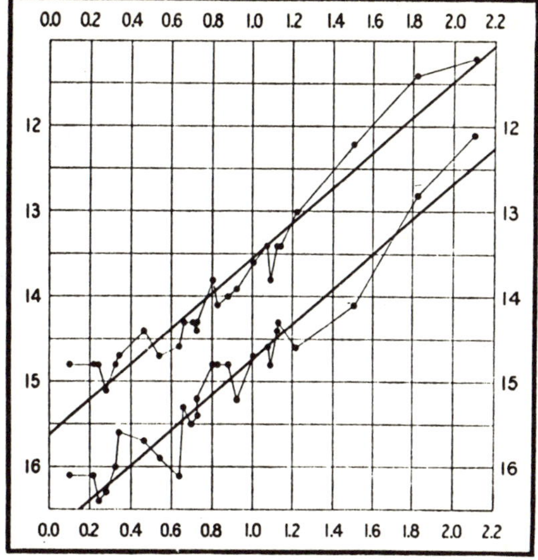

69 *A diagram illustrating the period-luminosity relationship of Cepheid variables, used by Henrietta Leavitt in her paper in* Harvard Circular, *no 173, 1912 (see extract 48)*

and found in it a Cepheid variable type of star, then, by measuring the apparent brightness of the star and its period, its true brightness could be calculated. From this, the distance of the globular cluster could be determined. Again, by observing the distribution of globular clusters, and other star clusters and stars, the shape of the star system could be determined. Shapley worked this out in as much detail as he could, using distances measured in kiloparsecs, 1 kiloparsec being equal to rather more than 3¼ million light-years (1 light-year is the distance light travels in a year, or approximately six million million miles):

The attempt to find the extent of the Galaxy from measures of its star clusters succeeds in the gross but not in detail. The clusters dimly outline the size but show little of the structure. We may confidently expect that analyses of star motions, star clouds, and individual stellar distances will in time reveal with more than present clarity the significance of our galactic system in the total material universe.

. . . . .

Minimum dimensions of the galactic system are found from the distances separating some of its stars and star clusters. Revising earlier values, we now find 70 kiloparsecs as the diameter in the galactic plane. The diameter perpendicular to the plane is probably about one tenth as great, with, however, a quantity of far outlying stars that are probably affiliated.

The center of the Galaxy, judged from the distribution of clusters and of several kinds of high-luminosity stars, is in the general direction of 17h 30m, −30° (Sagittarius) and approximately 16 kiloparsecs distant from the sun. Around this massive nucleus the neighboring galactic stars appear to move with a velocity of some 300 kilometers a second and a period well in excess of 100,000,000 years.

Modern investigations have modified these figures slightly—the diameter of the Galaxy and the distance of the Sun from the centre are rather smaller than Shapley estimated, yet there has been little change in the position of the centre.

50 Jan Oort on rotation of the galaxy
But is the Galaxy in motion? Kapteyn's two star stream theory made it appear to have motions that might best be interpreted as rotation, and in 1926 Bertil Lindblad (1895-1965) suggested that the Milky Way system of stars was really a collection of small flattened oval systems of different kinds of stars, all in rotation about the same centre, and a year later Jan Oort found that the Galaxy was rotating at different rates when velocities were

In order to explain the rotation there must be near the centre an attracting mass of at least 8×10^{10} times the mass of the sun. There remains the difficulty why we do not observe this large mass. Near 6,000 parsecs Kapteyn and van Rhijn [Pieter van Rhijn (1886-1960)] find an almost negligible density, whereas it *should* be very much greater than in our neighbourhood. Part of the discrepancy may have resulted from the approximative character of their solution, in which all galactic longitudes were combined. Discussing various galactic regions separately Kreiken [E. A. Kreiken of The Astronomical Institute of the University of Ankara] finds indications of a centre near 314° longitude, at a distance of 2,270 parsecs which is in the right direction, but certainly at too small a distance and too little defined. The most probable explanation is that the decrease of density in the galactic plane indicated for larger distances is mainly due to obscuration by dark matter. Such a hypothesis receives considerable support from the marked avoidance of the galactic plane by the globular clusters, a phenomenon for which up to the present time no other well defensible explanation has been put forward.

. . . . .

It has been shown from radial velocities that for all distant galactic objects there exist systematic motions varying with the galactic longitudes of the stars considered. The relative systematic motions are always the same nature and they increase roughly proportional with the distance of the objects. Probably the simplest explanation is that of non-uniform rotation of the galactic system around a very distant centre. This explanation is capable of representing all the observed systematic motions within their range of uncertainty (except perhaps in the case of the *B* stars). If with this supposition we compute the position of the centre from the radial

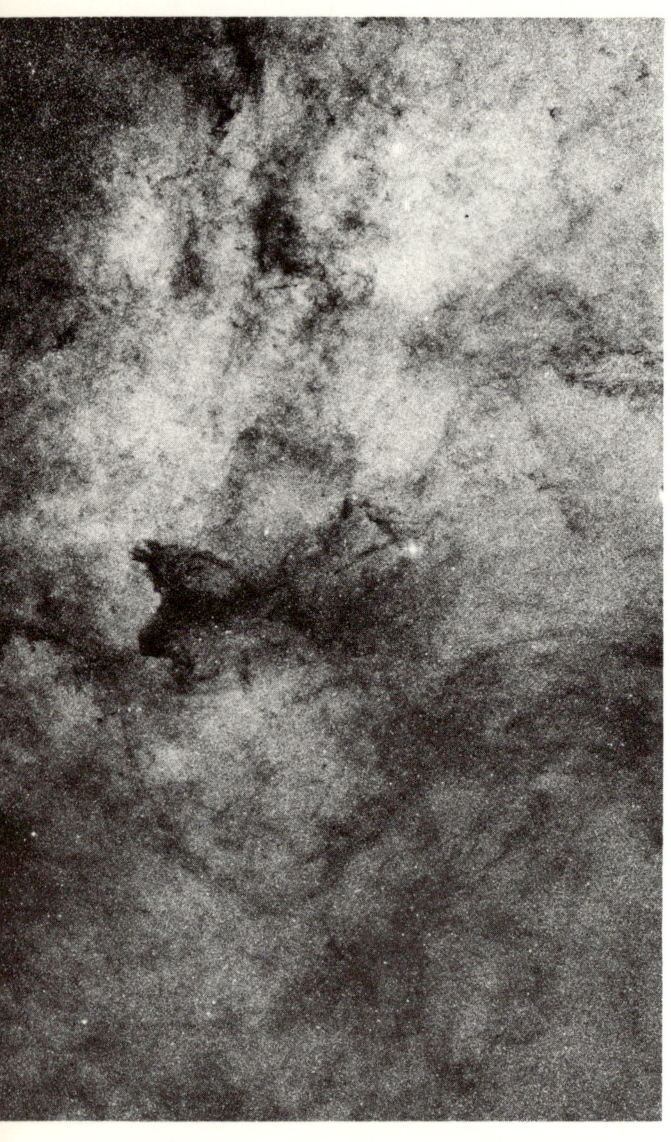

70 *The Milky Way between the constellations Scutum and Scorpius. In the centre of the photograph lies the region of Sagittarius, that is the central region of our Galaxy. 'Lanes' of dark matter, referred to by Oort (extract 50), can clearly be seen*

velocities, we find that it lies in the galactic plane, either at 323° longitude or at the opposite point. The first direction is in remarkably close agreement with the longitude of the centre of the system of globular clusters (325°). The observations would therefore seem to confirm LINDBLAD's hypothesis of a rotation of the entire galactic system around the latter centre.

51 Reber on radio radiation from space

Observationally, the difficulties in deciding the shape of the Galaxy were due to the presence of dust, gas and stars lying around the central regions. This filtered the starlight so that there was a limit to what could be seen. However, the Galaxy emitted not only light. In the early nineteenth century it became known that rays were emitted beyond the violet and the red, but in 1936 a radio engineer Karl Jansky (1905-50) discovered radio radiation from the Sagittarius section of the Milky Way (figure 71). This was followed up by Grote Reber, an American amateur radio experimenter, who confirmed these results and made other tests:

A few bright stars, such as Vega, Sirius, Antares,

71 *The directional antenna used by Karl Jansky whose observations led in due course to the development of radio astronomy*

Deneb, and the Sun, gave negative results. Mars and the Orion nebula also gave no readable indication. If radiation is present from any of these objects, the intensity is below 10^{-25} watts per square centimeter per circular degree per kilocycle band width at 162 megacycles. The only other positive results are from the great nebula in Andromeda, with a mean of four readings, giving a maximum intensity of 8×10^{-26} watts per square centimeter per circular degree per kilocycle band width. The foregoing observations confirm previous evidence [see figure 72] that radiation in the radio spectrum is apparently coming from the direction of the Milky Way. The intensity is a function of galactic longitude.

In the Netherlands, during World War II, Hendrik Van de Hulst worked out that, theoretically, a special radio aerial and receiver—a radio telescope—should be able to detect cold hydrogen gas in space, even though it emitted no optical radiation and was therefore

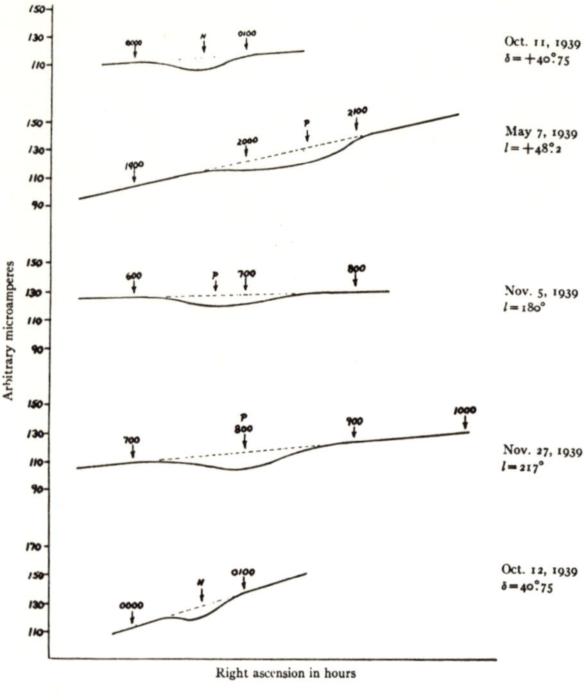

72 *A diagram that Reber used in his paper on cosmic radio radiation in the* Astrophysical Journal *in 1940. It shows his typical observational records; the letter P denoting the position of the plane of the Milky Way, and N the centre of the Andromeda galaxy (M31)*

invisible in an ordinary telescope. When the war was over and suitable radio telescopes were available, the theory was put to the test and the gas was detected. Mapping its presence and using the evidence of optical observations and other radio observations at various wavelengths, it was found that our Galaxy was not only disk-shaped, but also possessed spiral arms (figure 73).

52 Hubble and Humason on the movement of Galaxies

By the time this knowledge came, optical astronomers had successfully tackled the question of the spiral

90

nebulae and other more concentrated, elliptically shaped nebulae that also gave star-like spectra. Between 1919 and 1926 Edwin Hubble at Mount Wilson Observatory managed to resolve separate stars in the large spiral nebula in Andromeda (M 31) and a spiral in Triangulum (M 33), and to detect some Cepheid variables in them. Calculations showed that M 31 and M 33 lay at a distance of some 870,000 light-years, and although these figures had later to be revised, it became clear that each was a separate star island—an independent Milky Way system. With thousands of such objects known, it was realised that the universe was composed of myriads of such separate extra-galactic nebulae or 'galaxies', extending as far as the telescope could probe.

But the existence of separate galaxies was not all. Observations made between 1916 and 1917 showed that every galaxy displayed a shift in its spectral lines, and that in almost every case the shift was towards the red end of the spectrum. The galaxies were all moving outwards, away from each other, just as if the whole universe was expanding. And their velocities were not fixed—there was a distance-velocity relationship that showed the more distant to be moving more quickly:

Abstract. Methods of determining distances of extra-galactic nebulae are discussed, and the mean absolute magnitude is revised on the basis of (1) Shapley's revision of the zero-point of the period-luminosity curve for Cepheids, and (2) more extensive observations of stars involved in nebulae. The revised value is M(vis) [visual magnitude] = −14.9.

......

The velocity-distance relation is re-examined with the aid of 40 new velocities, 26 of which refer to nebulae in 8 clusters or groups. Distances of the clusters, ranging out to about 32 million parsecs [1 parsec=3.26 light-years], have been derived from the most frequent apparent magnitudes. The velocity displacements reduce the apparent magnitudes by amounts which become appreci-

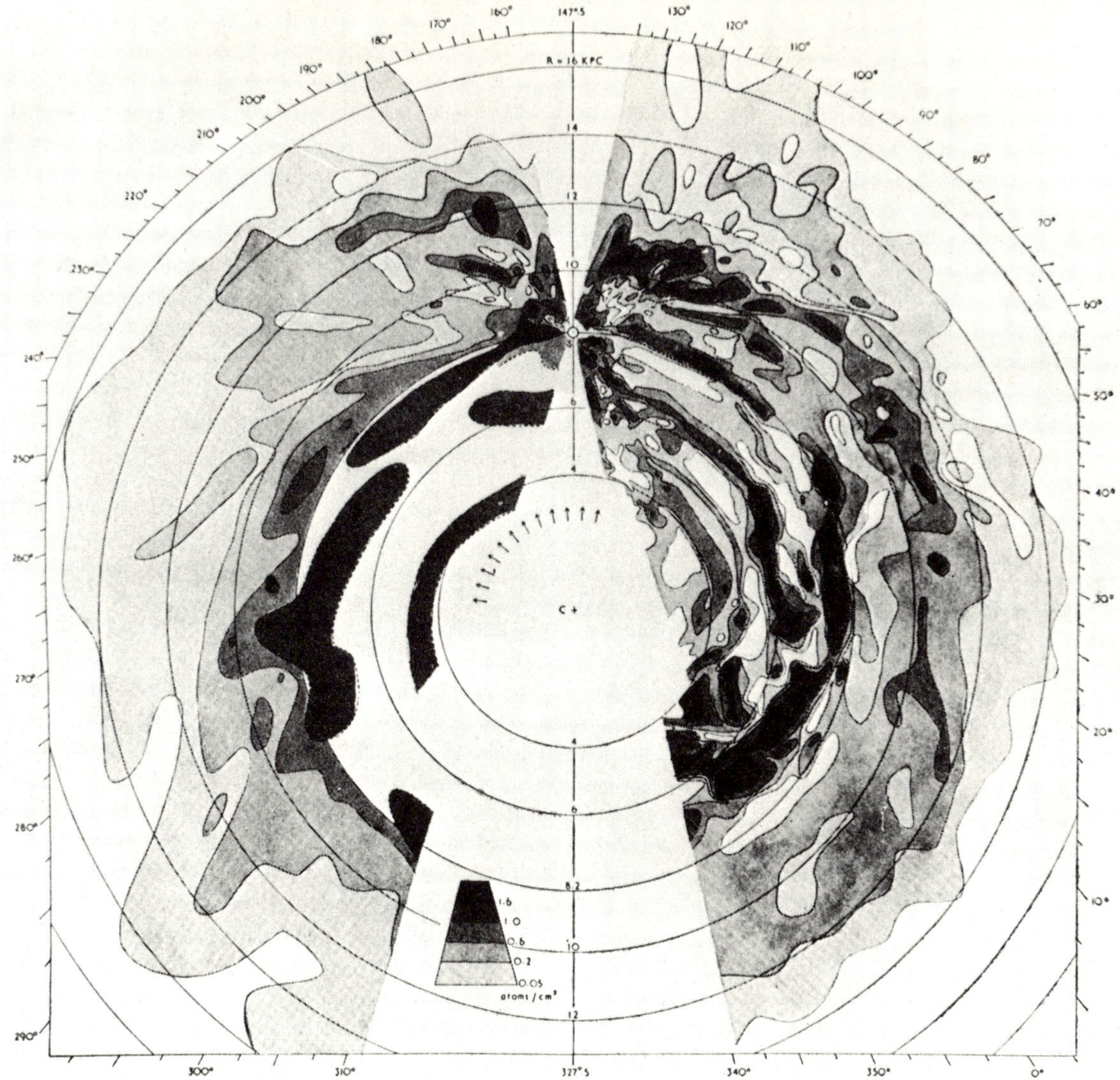

able for the most distant clusters.

The new data extend out to about eighteen times the distance available in the first formulation of the velocity-distance relation, but the form of the relation remains unchanged except for the revision of the unit of distance. The relation is

$$\text{Vel.} = \frac{\text{Dist. (parsecs)}}{1790},$$

and the uncertainty is estimated to be of the order of 10 per cent.

Further observations showed that many galaxies appeared also to be linked in clusters, but it was in the

73 *A radio map of hydrogen in the Galaxy, from a report on the spiral nature of the galactic system by Jan Oort, F. T. Kerr and G. Westerhout, prepared in 1958 for the Royal Astronomical Society. The diagram was drawn using results obtained in Leyden and in Sydney. One spiral arm— the Perseus arm—can readily be seen at a distance of 10·5 kiloparsecs from longitudes 65° to 130°. From* Monthly Notices of the Royal Astronomical Society, *volume 118, plate 6*

1950s when radio telescopes had been further developed (figure 74) that other evidence began to accumulate. Radio radiation was detected from discrete objects scattered about all over the sky, and from objects lying close to the plane of the Milky Way. The latter were taken as objects within our Galaxy—nebulae, dark hydrogen gas, and so on—and the rest were thought perhaps to be from sources far out in space. Radio radiation was detected from the Andromeda galaxy and some other galaxies near at hand, but it became increasingly important to see if an identification could be made between the majority of the sources and visual objects. The difficulty was one of determining the positions of the radio sources precisely enough—Sir Martin Ryle's early large 'interferometer' radio telescope at Cambridge only gave positions correct to a few minutes of arc, whereas photographs resolved objects a few seconds from each other. But by 1951 special interferometer techniques (figure 81) had reduced some of the areas of uncertainty:

74 *The 250ft diameter radio telescope at Jodrell Bank, Cheshire, one of the two largest steerable radio telescopes in the world*

53 Baade and Minkowski on optical identification of radio sources

Only very few individual sources of cosmic radio emission have been identified with conspicuous astronomical objects. Although the sources in Cassiopeia and Cygnus A are among the brightest and earliest-known radio sources in the sky, all attempts to identify them with astronomical objects in the visible range have failed so far. In the fall of 1951, F. G. Smith, of the Cavendish Laboratory, communicated to us in advance of publication new positions for both sources which were very much more accurate than any previous data. Using an interferometric method, Smith had reduced the uncertainties in the positions to $\pm 1^s$ in right ascension and $\pm 40''$ in declination. As it turned out, the new positions were accurate enough for an unambiguous identification of both radio sources on plates taken in September, 1951, at the 200-inch telescope. The Cassiopeia source coincides with a galactic-emission

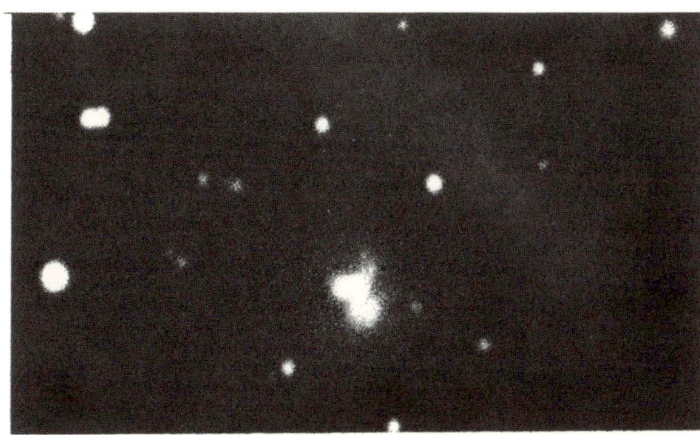

75 *The radio source 'Cygnus A' that, on optical examination, shows two galaxies in contact. They have been variously interpreted as being in collision or as splitting apart, and clearly appear to be galaxies—perhaps spirals with closed arms (type SO?)—interpenetrating one another*

nebulosity of a new type, whereas Cygnus A is an extragalactic affair, two galaxies in collision. Quite similar to Cassiopeia is one of the fainter radio sources, Puppis A.

54 Dewhirst on optical identifications of radio sources

In 1955 the position had further improved; Ryle had devised a new type of radio interferometer telescope, and weaker sources were observed. In 1959 and 1962 this work was extended, but in 1959 David Dewhirst was able to provide a bigger list of identifications, and it seemed that the sources were mostly either peculiar objects or the elliptical type of galaxy:

Previous attempts to identify any large proportion of the discrete sources discovered at meter wavelengths have met with small success. In the investigation previously reported here an extensive search has been made on the original plates of the 48-inch Palomar—National Geographic Society Sky Survey, using the available published radio data, but more especially the as yet unpublished results of a survey between +50 and −10 degrees declination that has been made with the interferometer of the Mullard Radio Astronomy Observatory, Cambridge. . . .

All of the galaxies designated [in a list not given here] as double or multiple E . . . require further investigation. On the scale of the Sky Survey plates it is difficult to distinguish between E and So [elliptical and closed type spirals] galaxies, but it seems certain that all the systems lack spiral arms or dust. The most striking result of the investigation, if the identifications are tenable, is that we recognize for the first time a new type of object which tends to be a source, namely close double or multiple E galaxies, usually sufficiently close to share a common outer extended envelope, and occurring either as the brightest members of small groups of E or So galaxies (as distinct from rich extended clusters), or as field objects. It is tempting to suppose, pending the collection of further spectroscopic information, that we are dealing with radio emission associated with a particular phase in the development of small groups of E and So galaxies. The systems appear to be physically distinguishable from such 'freak' objects as NGC 5128 and NGC 1275, characterized by their rich emission spectra and the presence of irregular dust patches.

55 Hewish and others on the discovery of the first 'pulsars'

Optical identifications have increased along with the accuracy of radio positions, and one important development (the discovery of 'quasars') will come more conveniently in the next section. However, in February 1968 another discovery was announced:

In July 1967, a large radio telescope operating at a frequency of 81.5 MHz was brought into use at the Mullard Radio Astronomy Observatory. This instrument was designed to investigate the angular structure of compact radio sources by observing the scintillation caused by the irregular structure of the interplanetary medium. The initial survey includes the whole sky in

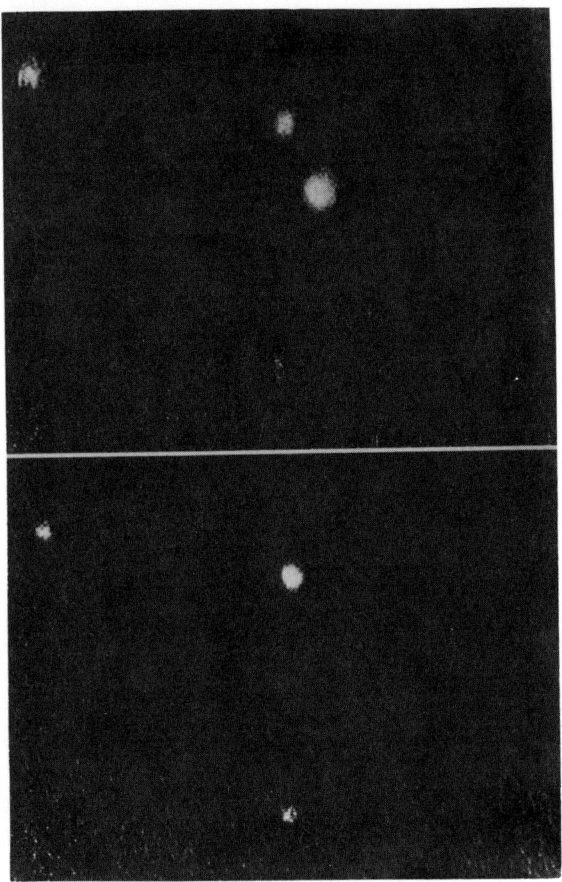

76 *Photographs showing the pulsar (NP 0532) in the Crab nebula (M1). The two exposures show the pulsar radiating and not radiating at visible wavelengths. These were taken using a television camera on the telescope and special electronic techniques to capture the light pulsations when the pulsar is both 'on' and 'off'. The pulsar period is 0·033 seconds. Photographed on 1969 February 3*

sporadic interference were repeatedly observed at a fixed declination and right ascension: this result showed that the source could not be terrestrial in origin.

Systematic investigations were started in November and high speed records showed that the signals, when present, consisted of a series of pulses each lasting $\sim$0.3 s and with a repetition period of about 1.337 s which was soon found to be maintained with extreme accuracy. Further observations have shown that the true period is constant to better than 1 part in 10^7 although there is a systematic variation which can be ascribed to the orbital motion of the Earth. The impulsive nature of the recorded signals is caused by the periodic passage of a signal of descending frequency through the 1 MHz pass band of the receiver.

The remarkable nature of these signals at first suggested an origin in terms of man-made transmission which might arise from deep space probes, planetary radar or the reflexion of terrestrial signals from the Moon. None of these interpretations can, however, be accepted because the absence of any parallax shows that the source lies far outside the solar system. A preliminary search for further pulsating sources has already revealed the presence of three others having remarkably similar properties which suggests that this type of source may be relatively common at a low flux density. A tentative explanation of these unusual sources in terms of the stable oscillations of white dwarf or neutron stars is proposed.

The nature of pulsars, which lie within our own galaxy, is not yet certain, although it seems not unlikely that they are 'neutron stars'—stars in collapsed states, at the ends of their lives. But whatever their true nature, astronomers are using them as radio beacons to give information on magnetic fields in space, and to aid a still fuller understanding of the layout of our galaxy and thus of other galaxies.

the declination range $-08° < \delta < 44°$ and this area is scanned once a week. A large fraction of the sky is thus under regular surveillance. Soon after the instrument was brought into operation it was noticed that signals which appeared at first to be weak

CHAPTER SIX

Beginning and End

The question of how the Earth was formed, the problems that arise when we consider the beginning of the universe and its ultimate end, may, in essence, be unanswerable. Even so this does not prevent attempts at informed speculation, especially since these are questions that are bound to be asked by inquiring minds of every age. Aristotle talked of a cyclic universe that was eternal; the philosophers of western Christendom of a divine creation out of nothing by an eternal God, and an original chaos that changed into order by divine intervention.

56 Kant on formation of the sun and planets

Only in the eighteenth century was there enough evidence to begin considering in detail the possibilities of a material creation. In days in which knowledge was not so specialised, Immanuel Kant (1724-1804) discussed the formation of the planets in his ALLGEMEINE NATURGESICHTE UND THEORIE DES HIMMELS (*General Natural History and Theory of the Heavens*):

In the present constitution of space, in which the globes of the whole planetary world revolve, there exists no material cause which could impress or direct their movements. This space is completely empty, or, at least, as good as empty; it must therefore have formerly been in another condition, and filled with enough of potential matter capable of transmitting motion to all the heavenly bodies found in it and bringing them into accordance with its own motion, so as to make them all concordant with each other. . . .

I assume that all the material of which the globes belonging to our solar system—all the planets and comets—consist, at the beginning of all things was decomposed into its primary elements, and filled the whole space of the universe in which the bodies formed out of it now revolve. This state of nature, when viewed in and by itself without any reference to a system, seems to be the very simplest that can follow upon nothing. . . .

77 *A cluster of galaxies in the constellation Leo, with an elliptical galaxy NGC 3193 (top left); a normal spiral NGC 3190 (bottom right); and barred spirals NGC 3185 (upper left centre) and NGC 3187 (upper right). The distance of this cluster is of the order of 500 million light-years. At the time of Kant and Laplace, only our own Galaxy was known, and that but very imperfectly*

From what has been said, it will appear that if a point is situated in a very large space where the attraction of the elements there situated acts more strongly than elsewhere, then the matter of the elementary particles scattered throughout the whole region will fall to that point. The first effect of this general fall is the formation of a body at its centre of attraction which, so to speak, grows from an infinitely small nucleus by rapid strides; and in the proportion in which this mass increases, it also draws with greater force the surrounding particles to unite with it. . . .

If we, therefore, consider this revolving elementary matter of the world in that state into which it puts itself by attraction and by the mechanical consequence of the general laws of resistance, we see a region of space extending from the centre of the sun to unknown distances, contained between two planes not far distant from each other, in the middle of which the general plane of reference is situated. And this elementary matter is diffused in this space within which all the contained particles—each according to the proportion of its distance and of the attraction which prevails there—perform regulated circular movements in free revolutions.

57 Laplace on the formation of the solar system

Kant had been stimulated to advance his ideas after reading Thomas Wright's conjectures about the layout of the universe (Chapter 5), and they were followed up by Pierre Laplace (1749-1827) who, in 1796, suggested that the Sun and the solar system originated from a large primordial nebula:

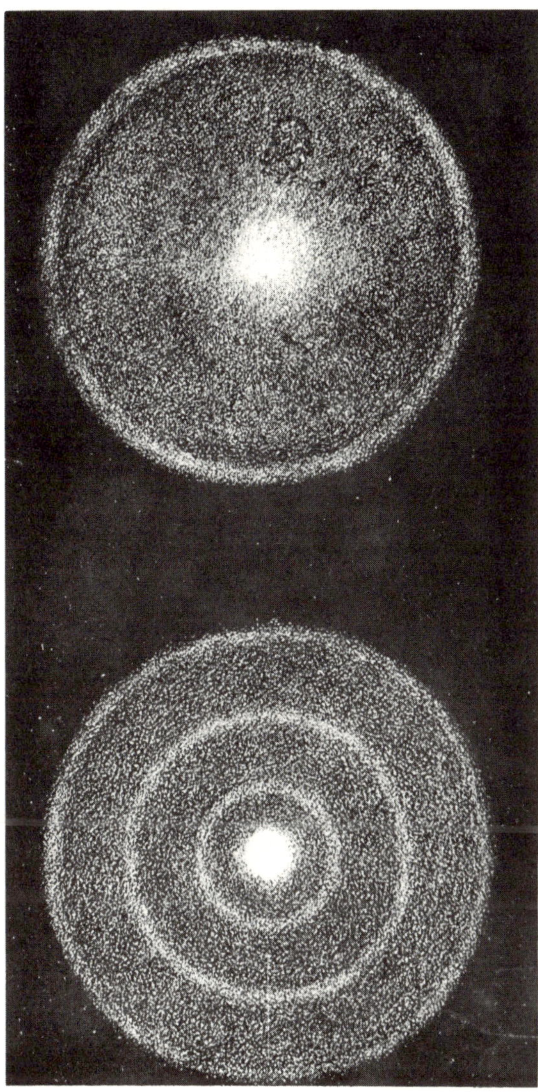

78 *Laplace's nebular hypothesis, as illustrated in*
J. P. Nichol, Views of the Architecture of the Heavens
(London 1843). In the upper half of the engraving, the
nebula is shown with a single outer ring; in the lower half
the nucleus is more concentrated and three rings of
condensing matter have formed

Thus to investigate the cause of the primitive motion of the planets, we have given the five following phenomena: 1st, The motions of planets in the same direction, and nearly in the same plane. 2d, The motion of their satellites in the same direction, and nearly in the same plane with those of the planets. 3d, The motion of rotation of these different bodies, and of the Sun in the same direction as their motion of projection, and in planes but little different. 4th, The small excentricity of the orbits of the planets, and of their satellites. 5th, The great excentricity of the orbits of comets, although their inclinations may have been left to chance.

. . . . . .

. . . Let us see if it is possible to arrive at their true cause.

Whatever be its nature, since it has produced or directed the motion of the planets and their satellites, it must have embraced all these bodies, and considering the prodigious distance which separates them, they can only be a fluid of immense extent. To have given in the same direction, a motion nearly circular round the Sun, this fluid must have surrounded the luminary like an atmosphere. This view, therefore, of planetary motion, leads us to think, that in consequence of excessive heat, the atmosphere of the Sun originally extended beyond the orbits of all the planets, and that it has gradually contracted itself to its present limits, which may have taken place from causes similar to those which caused the famous star that suddenly appeared in 1572, in the constellation Cassiopæa, to shine with the most brilliant splendour during many months.

The great excentricity of the orbits of comets, leads to the same result; . . .

But how has it determined the motions of revolution and rotation of the planets? If these bodies had penetrated this fluid, its resistance would have caused them to fall into the Sun. We may then conjecture, that they have been formed at the successive bounds of this atmosphere, by the condensation of zones, which it must have abandoned in the plane of its equator, and in becoming cold have condensed themselves towards the surface of this luminary, as we have seen in the preceding Book. One may likewise conjecture, that the satellites have been formed in a similar way by the atmosphere of the planets. The five phenomena, explained above, naturally result from this hypothesis, to which the rings of Saturn add an additional degree of probability.

58 Jeans on the tidal theory of planetary formation

As time went on, and the mathematics of Laplace's theory were studied in detail, it became clear that it would not work in the way he had proposed. An alternative hypothesis was needed, and in 1906 in Chicago Thomas Chamberlin (1843-1928) and Forest Moulton (1872-1952) suggested a quite different idea. They considered it possible that the planets could have been formed from material drawn out of the Sun by another star that passed close to it. In 1916 James Jeans (1877-1946) developed this into his tidal theory:

TIDAL THEORY. Actual collisions must be so exceedingly rare that we can leave them out of account. When two stars pass close to one another without collision, the primary effect must be that each raises tides in the other. The closer the approach, the higher the tides in general, although something must depend also on the speed with which the bodies pass one another, because this determines the length of time during which they influence one another.

. . . . . .

If the approach is very close indeed, the tides may assume an entirely different aspect from the feeble tides which the sun and moon raise in our oceans; they may take the exaggerated forms of high mountains of matter moving over the surface of the star. An even closer approach may transform these mountains into long arms of gas drawn out from the body of the star. If, as will generally be the case, the two stars are of unequal weights, the lesser will in general suffer more disturbance than the weightier.

THE BIRTH OF PLANETS. The long arm or filament of matter drawn out of a star by tidal action is at first continuous in its structure, but analysis shews that it provides a fit subject for the operation of what we have called "Gravitational Instability." Condensations begin to form in this long arm of gas, in the way already described. As before, the smaller condensations are dissipated, while the larger increase in intensity until finally the filament breaks up into a number of detached masses—planets have been born out of the smaller star.

As the years passed, variations of this tidal hypothesis were proposed. In 1934 H. N. Russell suggested that the Sun might have been a member of a binary system, whose companion was carried away by a passing star, the planets forming from material torn off the companion. In 1941 Raymond Lyttleton disposed of this idea, but pointed out that all might be well if the Sun was once part of a triple star system, and the system swept up dust and gas from space as it moved in orbit around the centre of the Galaxy. But gradually it became clear that since all stars were primarily composed of hydrogen, there was difficulty in accounting for the abundance of heavy atoms found in the planets. Two courses seemed likely. One could use the tidal hypothesis, return to a binary system, and suggest, as Fred Hoyle did in 1944, that the companion star was a super-nova that exploded. In this case the supernova would fly off but leave behind some of the heavy atoms found at the time of the explosion. The other possibility was to return to Laplace and modify the nebular hypothesis.

59 Hoyle on the solar nebula

It was C. F. von Weizsäcker who, in 1945, went back to a solar nebula, with the Sun forming at the centre. He showed that instead of breaking up into rings as Laplace

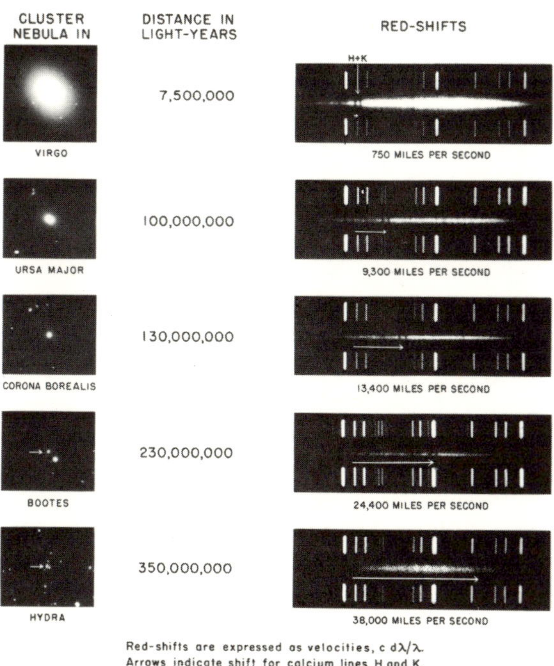

79 *The relationship between the red-shifts of galaxies and their distances*

thought, it would form into oval eddies, and from these the planets could be condensed. Gerard Kuiper developed the hypothesis, suggesting that the eddies would first form 'protoplanets' which only later became planets. Meanwhile others considered the capture of dust and gas by the Sun, and the effects of magnetism as well as gravitation on the material. Fred Hoyle, with William Fowler, has recently proposed that heavier atoms were generated by atomic particles emitted from a primeval Sun, and Hoyle has also considered the behaviour of the solar nebula, discussing its rotational power (angular momentum) and the effect of magnetism on the electrified gases there that would act like a fluid (hydromagnetic behaviour):

A consideration of star formation from the interstellar gas shows that the primitive solar condensation was probably endowed with angular momentum very much greater than that now possessed by the Sun. It is found, however, that the initial angular momentum agrees with that possessed by the primeval planetary material. It is suggested therefore that the angular momentum of the solar nebula was transferred to the planetary material. A similar process of transfer appears to take place generally in dwarf stars. Massive early-type stars, on the other hand, possess the expected angular momentum, and presumably do not have planetary systems. In the solar case, the process of angular momentum transfer was probably hydromagnetic—a field with intensity $\sim$I gauss being required to connect the Sun and the planetary material [A gauss is a measure of magnetic intensity: the Earth's general magnetic field is a little under 0.5 gauss].

As the planetary material acquired angular momentum it moved farther and farther away from the solar condensation. Materials of low volatility condensed out of the gas and were left behind as the main mass of gas moved outwards. The fact that the terrestrial planets are (i) of small mass (ii) composed almost wholly of low volatility materials (iii) on the inside of the system, is attributed to this leaving behind of the low volatility materials.

60 Eddington on Einstein's theory of relativity

The question of the origin of the universe is even more unsettled than the origin of the solar system. Even before the discovery of the galaxies, the origin of the universe took on a new aspect with the advent in 1905 and 1915 of Einstein's papers on the theory of relativity, a theory that met with considerable resistance initially, but was gradually accepted due in no small degree to the

By his theory of relativity Albert Einstein has provoked a revolution of thought in physical science.

The achievement consists essentially in this:—Einstein has succeeded in separating far more completely than hitherto the share of the observer and the share of external nature in the things we see happen. The perception of an object by an observer depends on his own situation and circumstances; for example, distance will make it appear smaller and dimmer. We make allowance for this almost unconsciously in interpreting what we see. But it now appears that the allowance made for the *motion* of the observer has hitherto been too crude—a fact overlooked because in practice all observers share nearly the same motion, that of the earth. Physical space and time are found to be closely bound up with this motion of the observer; and only an amorphous combination of the two is left inherent in the external world. When space and time are relegated to their proper source—the observer—the world of nature which remains appears strangely unfamiliar; but it is in reality simplified, and the underlying unity of the principal phenomena is now clearly revealed. The deductions from this new outlook have, with one doubtful exception, been confirmed when tested by experiment.

The theory led to a number of mathematically plausible schemes for the universe, one due to Willem de Sitter (1872-1935) giving a universe that was expanding into space and obviously linked with the observation that the majority of galaxies possess a red-shift (Chapter 5):

61 Hubble on the expanding universe

The outstanding feature, however, is the possibility that the velocity-distance relation may represent the de Sitter effect [expanding universe], and hence that numerical data may be introduced into discussion of the general curvature of space. In the de Sitter cosmology, displacements of the spectra arise from two sources, an apparent slowing down to atomic vibrations and a general tendency of material particles to scatter. The latter involves an acceleration and hence introduces the element of time. The relative importance of these two effects should determine the form of the relation between distances and observed velocities; and in this connection it may be emphasized that the linear relation found in the present discussion is a first approximation representing a restricted range in distance.

62 Lemaître on the origin of the universe

Subsequent research has modified de Sitter's ideas, but there now seems little doubt about our being in an expanding universe. If then we agree that the universe is expanding, and begin to look back into time, galaxies should have been closer to one another than they are now. Taken to its extreme, the galaxies should have been packed tightly together when the universe began; and between 1934 and 1935 the French cosmologist, Georges Lemaitre used this as a basis for what has since become called the 'Big Bang' theory. Eddington believed that one could not discuss the creation of the universe in a way that was scientifically meaningful because it was a unique event, but Lemaître thought otherwise:

Sir Arthur Eddington states, that philosophically, the notion of a beginning of the present order of Nature is repugnant to him. I would rather be inclined to think that the present state of quantum theory [the theory that energy is 'atomic', existing only in discrete bits or 'quanta'] suggests a beginning of the world very different from the present order of Nature. Thermodynamic principles from the point of view of quantum theory may be stated as follows: (1) Energy of constant total amount is distributed in distinct quanta. (2) The number of distinct quanta is ever increasing. If we go back in the course of time we must find fewer and fewer quanta, until we find all the energy of the universe packed in a few or even in a unique quantum.

Now, in atomic processes, the notions of space and time are no more than statistical notions; they fade out when applied to individual phenomena involving but a small number of quanta. If the world began with a single quantum, the notions of space and time would altogether fail to have any meaning at the beginning; they would only begin to have a sensible meaning when the original quantum had been divided into a sufficient number of quanta. If this suggestion is correct, the beginning of the world happened a little before the beginning of space and time. I think that such a beginning of the world is far enough from the present order of Nature to be not at all repugnant.

It may be difficult to follow up the idea in detail as we are not yet able to count the quantum packets in every case. For example, it may be that an atomic nucleus must be counted as a unique quantum, the atomic number acting as a kind of quantum number. If the future development of quantum theory happens to turn in that direction, we could conceive the beginning of the universe in the form of a unique atom, the atomic weight of which is the total mass of the universe. This highly unstable atom would divide in smaller and smaller atoms by a kind of super-radioactive process. Some remnant of this process might, according to Sir James Jeans' idea, foster the heat of the stars until our low atomic number atoms allowed life to be possible.

Clearly the initial quantum could not conceal in itself the whole course of evolution; but, according to the principle of indeterminacy, that is not necessary. Our world is now understood to be a world where something really happens; the whole story of the world need not have been written down in the first quantum like a song on the disc of a phonograph. The whole matter of the world must have been present at the beginning, but the story it has to tell may be written step by step.

63 Gamow on the evolution of the universe

This theory was developed, notably by Lemaître himself, and by George Gamow in the United States who, in 1948, was prepared to discuss the details of the early stage:

The discovery of the red shift in the spectra of distant stellar galaxies revealed the important fact that our universe is in the state of uniform expansion, and raised an interesting question as to whether the present features of the universe could be understood as the result of its evolutionary development, which must have started a few thousand million years ago from a homogeneous state of extremely high density and temperature. We conclude first of all that the relative abundances of various atomic species (which were found to be essentially the same all over the observed region of the universe) must represent the most ancient archaeological document pertaining to the history of the universe. These abundances must have been established during the earliest stages of expansion when the temperature of the primordial matter was still sufficiently high to permit nuclear transformations to run through the entire range of chemical elements. It is also interesting to notice that the observed relative amounts of natural radioactive elements suggest that their nuclei must have been formed (presumably along with all other stable nuclei) rather soon after the beginning of the universal expansion. In fact, we notice that natural radioactive isotopes with the decay periods of many thousand million years (such as uranium-238, thorium-232 and samarium-148) are comparatively abundant, whereas those with decay periods measuring only several hundred million years are extremely rare (as uranium-235 and potassium-40). If, using the known decay periods and natural abundances of these isotopes, we try to calculate the date when they have been about as abundant as the corresponding isotopes of longer life, we find that it must have been a few thousand million years ago, in general agreement with the astronomically determined age of the universe.

64 Bondi and Gold on the steady-state theory of the expanding universe

But a new approach to the nature of the universe was made in 1948 by Herman Bondi and Thomas Gold, who suggested that, on a long-term view, the universe was always the same:

The applicability of the laws of terrestrial physics to cosmology is examined critically. It is found that terrestrial physics can be used unambiguously only in a stationary homogeneous universe. Therefore a strict logical basis for cosmology exists only in such a universe. The implications of assuming these properties is investigated.

Considerations of local thermodynamics show as clearly as astronomical observations that the universe must be expanding. Hence, there must be continuous creation of matter in space at a rate which is, however, far too low for direct observation. The observable properties of such an expanding stationary homogeneous universe are obtained, and all the observational tests are found to give good agreement.

The physical properties of the creation process are considered in some detail, and the possible formulation of a field theory [the appropriate mathematical equations expressing the gravitational fields in the universe] is critically discussed.

In the same volume of the Monthly Notices of the Royal Astronomical Society *as this paper appeared, Fred Hoyle published his 'New Model for the Expanding Universe'. Many astronomers were then wary of a mathematical term called the 'cosmical constant' which Einstein had introduced in the field equations because the expansion of the universe was not then known. Later he abandoned the term, but many other cosmologists, including Eddington and Lemaître, did not, even though the physical explanation of it was something of a mystery. Hoyle's model, similar to that of Bondi and Gold, used a steady-state universe with*

continuous creation, and removed the need for the constant.

65 Ryle on radio observations and the steady-state theory

Ever since 1948 the battle between the two basic kinds of theory has raged. In 1961 Ryle and his colleagues at

80 *The movable antennae of the world's first large 'aperture synthesis' interferometer type of radio telescope. An extension of the concept of the interferometer devised in 1890 by Albert Michelson (1852-1931), and used by him in its basic form on the 100-inch reflector at Mount Wilson Observatory, California, in 1920. This design of radio telescope by Sir Martin Ryle is equivalent, in its power to pinpoint the positions of radio sources, to a radio telescope with a dish equal in area to that covered by the distance over which this antenna moves, and the distance of the fixed antenna which the instrument also possesses*

Cambridge University believed they had evidence to show that the 'Big Bang' theories only were correct:

The results of some recent observations of radio sources have been used to investigate the 'luminosity function' [a measure of the radiation emitted by a source] for the sources. Without making any assumptions about their nature, it can be shown from radio observations alone that most of the sources lying in any given range of flux density are extra-galactic with an emission at 178 Mc/s which exceeds 10^{24} watts $(c/s)^{-1}$ ster^{-1}. [The energy emitted (10^{24} watts) per cycle at the frequency of observation $[(c/s)^{-1}]$ per steradian (one steradian is an area of a little over $59°$ by $59°$)]. If it is assumed either that the physical dimensions or that the optical luminosities of the sources are comparable with those of the Galaxy, then the emission must lie in the range $3 \times 10^{25} - 10^{27}$ watts $(c/s)^{-1}$ ster^{-1}.

From these figures it is possible to derive the expected number-flux density relationship according to different cosmological models and special consideration has been given to the predictions of the steady-state model.

With the new Cambridge interferometer it has become possible to observe sources considerably weaker than those reached in earlier surveys, and hence to make a more accurate determination of the actual number-flux density relationship; the new data has also allowed more detailed corrections for the effects of extended sources and source clustering to be made.

A comparison of these observational results with those predicted by the steady-state model shows a marked discrepancy, the number of sources observed with a flux density in the range 0.5 to 2×10^{-26} watts $(c/s)^{-1}$ m^{-2} [here measured per square metre (m^{-2}) of the aerial receiving area] being at least 3 ± 0.5 times that predicted by the model.

No attempt has been made to select an alternative model to account for the observations, but the results

81 *The 1,000ft diameter radio telescope of Cornell University, constructed in a natural bowl in the ground at Arecibo, Puerto Rico. Although it can only be steered a very few degrees (by altering the position of the antenna suspended above the reflecting bowl), the motion of the Earth permits different areas of the sky to be observed*

appear to provide conclusive evidence against the steady-state model.

The gist of this criticism is that at greater and greater distances, the number of sources per given volume of space increases. Since radio radiation, like light, takes time to travel, objects observed at greater and greater distances are observed as they were longer and longer ago. If then there is a crowding of distant sources, this means they were more concentrated in the past, a state of affairs to be expected on the Big Bang theory, but not on the steady-state where the universe appears the same on the average at every age.

66 Hoyle and Narlikar on the counting of distant radio sources

But Ryle's criticism did not long remain unchallenged:

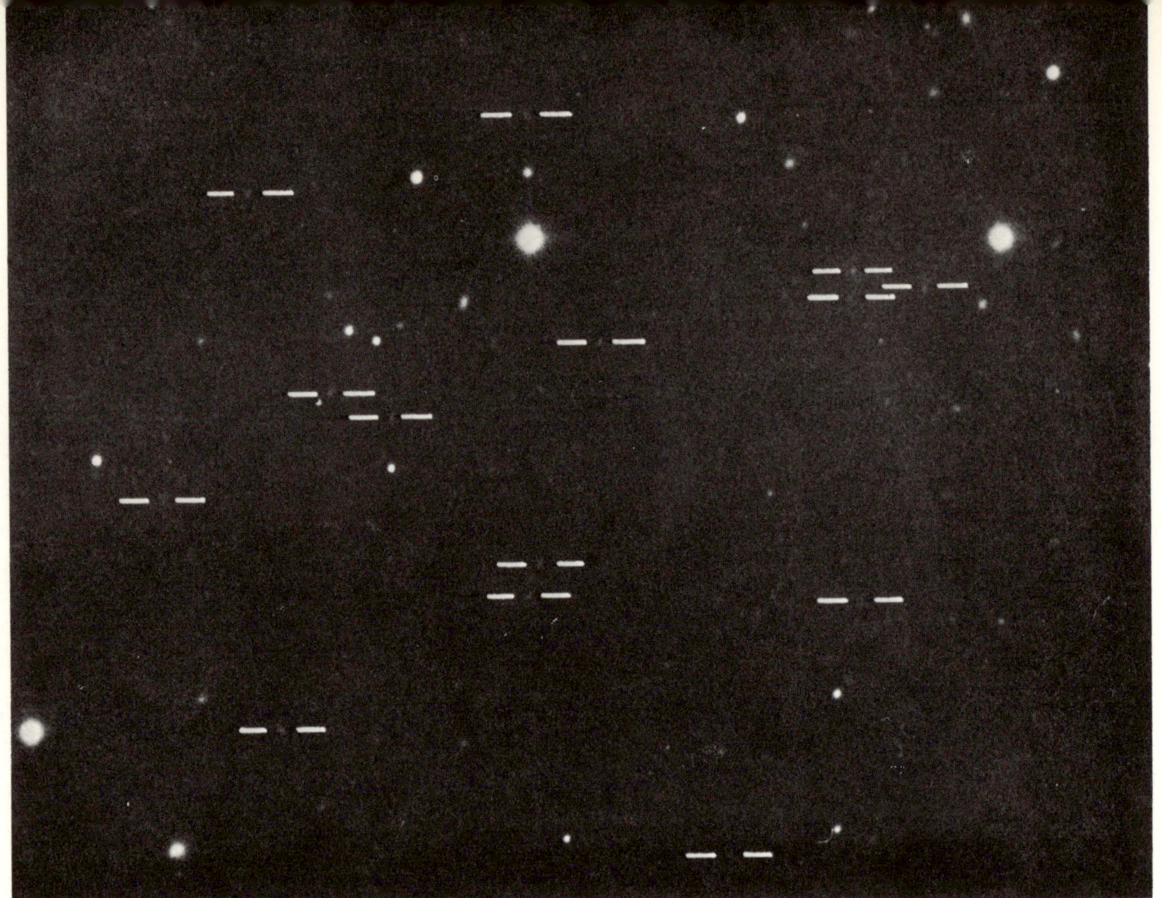

82 *Field of faint galaxies in the constellation Coma Berenices, some of which are indicated by the white dashes (the bright disks represent stars in our own Galaxy). These are very distant, lying at something of the order of 2,000 million to 4,000 million light-years. Photographed with the 200in telescope*

Hoyle and his colleague Jayant Narlikar were back in six months with an answer:

The problem of the number count of radio sources as a function of the incident flux is shown to depend on two crucial features: (i) the size and behaviour of condensations (ii) the dependence on age of the probability of a galaxy being a radio source. Provided the probability rises by a factor of $\sim 10^2$ for galaxies with ages from [a method of computing the age of galaxies from their speeds of recession and their distances. H^{-1} is an age of about 10^{10} years] H^{-1} to about $2.5H^{-1}$, provided the primary condensations of the steady-state theory possess initial dimensions of order ʒ0 megaparsecs, the radio source count can raise more steeply than is the case for sources uniformly distributed in Euclidean space [Euclidean space is the familiar kind of space used in everyday geometry. In relativity universes of today different kinds of space are adopted]. Each primary condensation contains of the order of 10^5 galaxies, which in the main expand apart from each other as the universe expands.

It is shown that a luminosity function can be chosen for the radio sources giving results consistent with observation, not only for the source count, but also with the data on angular diameters and with experience in the problem of optical identifications.

67 Dicke, Peebles, Roll and Wilkinson on cosmic black-body radiation

The steady-state theory was, then, slightly modified. But a new and serious blow came in 1965 when Arno Penzias and Robert Wilson, two American radio engineers, detected radiation from space at a wavelength of 7.35 centimetres, which Robert Dicke of Princeton thought should be detectable if the Big Bang theory were correct, or if the universe were an oscillating one—expanding as we now observe it to be, contracting later until compact once again, then expanding, and so on—which is another possible model derived from relativity theory. The radiation should come from the time when the universe was packed closely together and was like a giant ball of 'fire', which would cool as it expanded and should now be very cool indeed if the Big Bang or dense state occurred some ten thousand million years ago, as is widely believed. The radiation would be just like that obtained from a body that radiates throughout the entire spectrum, and which would be completely black when utterly cold—usually therefore referred to as a 'black body':

. . . it is of interest to inquire about the temperature of the universe in . . . earlier times. From this broader viewpoint we need not limit the discussion to closed oscillating models. Even if the universe had a singular origin it might have been extremely hot in the early stages.

Could the universe have been filled with black-body radiation from this possible high-temperature state? If so, it is important to notice that as the universe expands the cosmological redshift would serve adiabatically [without loss or gain of heat] to cool the radiation, while preserving the thermal character. The radiation temperature would vary inversely as the expansion parameter (radius) of the universe.

The presence of thermal radiation remaining from the fireball is to be expected if we can trace the expansion of the universe back to a time when the temperature was of the order of 10^{10}°K [degrees Kelvin: 0°K is 'absolute zero'—the lowest possible temperature]. . . .

. . . we recently learned that Penzias and Wilson (1965) of the Bell Telephone Laboratories have

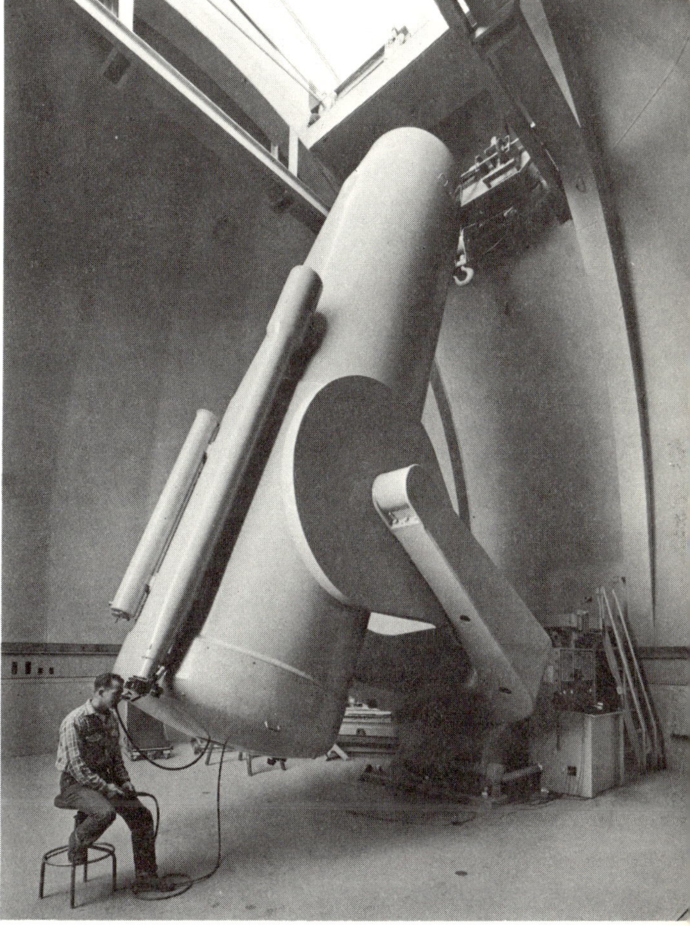

83 *The 48in Schmidt telescope at Palomar. This telescope, based on the design of Bernhard Schmidt (1879-1935), gives sharp photographs over a large area of the sky—some 25 square degrees compared with the area of something less than one square degree given by a large reflector of conventional design like the 200in. The Schmidt requires a glass correcting plate a little larger in diameter than the main mirror*

observed background radiation at 7.3-cm wavelength. In attempting to eliminate (or account for) every

84 *The special horn radio antenna used by Penzias and Wilson at Holmdel, New Jersey, that led to the detection of cosmic 'background' radio radiation (see extract 67)*

contribution to the noise seen at the output of their receiver, they ended with a residual of $3.5° \pm 1°K$. Apparently this could only be due to radiation of unknown origin entering the antenna.

. . . .

A temperature in excess of $10^{10}°K$ during the highly contracted phase of the universe is strongly implied by a present temperature of $3.5°K$ for black-body radiation.

But the matter still appears to be in the balance. It has not yet been possible to make all the observations necessary at the crucial wavelengths to confirm that the radiation is truly of 'black-body' type and, moreover, it has been suggested that the radiation could come from matter lying in between galaxies. Yet cosmic radiation is

not the only problem facing cosmologists and their theories.

68 Greenstein and Schmidt on quasars

In 1960 the radio source 3C48 (item 48 in the third Cambridge catalogue of radio sources) was identified as what appeared to be a small blue star, yet its radio radiation was far more intense than could come from any star. Looking like a star but not being one, it was dubbed a 'quasi-stellar radio source' or quasar for short. Other quasars were identified. Jesse Greenstein and Maarten Schmidt, using the 200 inch telescope, examined the spectra of 3C48 and 3C273 and found astoundingly large red-shifts:

The present paper deals with optical objects of *stellar* appearance that have been associated with radio sources. The first radio source so identified was 3C 48, for which Matthews, Bolton, Greenstein, Münch, and Sandage (1960) announced the stellar

106

appearance of the associated optical object. Subsequently, the radio sources 3C 196 and 3C 286 were identified with similar optical objects (Matthews and Sandage 1963), as was the source 3C 147 (Schmidt and Matthews 1964). The optical spectra of these four quasi-stellar objects appeared quite dissimilar, and no satisfactory identifications of the emission features could be obtained. The identification of the radio source 3C 273 with a bright object of stellar appearance provided a clue when it was found that its spectrum could be understood on the basis of an unexpectedly large redshift (Schmidt 1963). The spectrum of 3C 48, although of a rather different nature, could be explained by an even larger redshift (Greenstein and Matthews 1963).

The present discussion is limited to the *quasi-stellar radio sources* 3C 48 and 3C 273; the observational data are given in Sections II and III. The possibility of interpreting the red-shift as the gravitational effect of either very dense or very massive objects is discussed in Section IV. The finally adopted interpretation of these quasi-stellar radio sources as distant, super-luminous objects in galaxies, or intergalactic objects, is discussed in the remaining sections.

Relativity does predict a red-shift caused by gravity, as well as the usual one due to motion in the line-of-sight. Ordinarily the gravitational shift is too small to bother about, but it was suggested as a cause for the quasar red-shifts as an alternative to their being very distant objects, since if they were very distant they would show an early stage of the universe's development, and militate against the steady-state theory.

69 McCrea on continual creation

Observations made since 1964 have shown that some quasars radiate little radio radiation, and that some have jets of matter associated with them (figure 87). But jets of matter are associated with some galaxies (figure 88), and even exploding galaxies are known. In view of all this, William McCrea's suggestion, made in

85 *The elliptical galaxy M87 (NGC 4486), showing a jet of material being ejected. Observations in visible, ultra-violet, X-ray, radio and infra-red wavelengths have been made of this object with a view to the fullest investigation of the phenomena involved*

1964, of the continual creation of matter within galaxies, instead of evenly throughout space as in the continuous creation of Bondi, Gold and Hoyle, is certainly worth quoting:

It is suggested that considerable clarification in cosmology results from the hypothesis that continual creation of new matter is a property of existing matter depending upon its physical state. Accordingly, it is envisaged that effectively all matter is in galaxies and that continual creation simply promotes the growth of galaxies; any galaxy, however, may occasionally eject a fragment of itself that becomes the embryo for the growth of a new galaxy. A steady-state model universe may be constructed in conformity with this picture. The proposed view appears to be in better accord with general physical concepts than that of current steady-state cosmology, where the rate of continual creation is taken to be effectively uniform in space and time and independent of already-existing matter. The new suggestion simplifies the problem of galaxy-formation; in requiring no important quantity of inter-galactic matter, it does not conflict with available observational evidence. If

86 *Twentieth-century observational astronomy has been revolutionised by the advent of radio astronomy and by the new techniques of space astronomy that permit observations to be made beyond the restrictive blanket of the Earth's atmosphere in far ultra-violet and the x-ray and γ-ray regions. An orbiting astronomical observatory (OAO) is here being prepared for its space role, and has been successfully launched since this photograph was taken*

the suggestion could be given a more quantitative formulation, it would apparently show whether we should see a universe composed of ordinary matter, or one composed of galaxies made of ordinary matter intermingled with galaxies made of anti-matter [atoms similar to those with which we are familiar but with opposite electric charges. On meeting our atoms there would be an annihilation of both kinds of matter], without demanding any special mechanism for separating matter and anti-matter. By discarding the hypothesis of a rate of creation that is a universal constant, cosmology based on continual creation becomes a more flexible concept; it may accommodate the author's suggestions regarding uncertainty in cosmology.

The true nature of quasars, whether the cosmic background radiation is of 'black-body' type, and the reasons for it, are still problems awaiting definitive solutions. Probably new observations will bring such surprises as will call for a radical revision of all current theories, perhaps forging a closer link between the sub-atomic world of the very small and the vastness of the astronomer's universe.

Only the future will tell.

List of Sources

CHAPTER I THE PLANETARY SYSTEM

1 Aristotle, *De Caelo*, 296ª. Translated by J. L. Stocks. Clarendon Press, Oxford, 1930
2 Ptolemy, *Almagest*, Bk I, section 3. Translated by R. Catesby Taliaferro. 'Great Books of the Western World', vol 16, Encyclopaedia Britannica, Chicago and London, 1952
3 Ptolemy, *Almagest*, Bk III, section 3. Translation same source as extract 2 above
4 Copernicus, *De Revolutionibus Orbium Coelestium*, Bk I, section 8. Translated by Charles Glenn Wallis. 'Great Books of the Western World', vol 16, Encyclopaedia Britannica, Chicago and London, 1952
5 Copernicus, *De Revolutionibus Orbium Coelestium*, Bk I, section 10. Translation same source as extract 4 above
6 Kepler, *Epitome astronomiæ Copernicanæ*, Bk V, part I. Translated by Charles Glenn Wallis. 'Great Books of the Western World', vol 16, Encyclopaedia Britannica, Chicago and London, 1952
7 Kepler, *Epitome astronomiæ Copernicanæ*, Bk V, part I. Translation same source as extract 6 above
8 Kepler, *Epitome astronomiæ Copernicanæ*, Bk IV, section 3. Translation same source as extract 6 above
9 Galileo, *Discorsi . . . à due nuove scienze*. Translated by Henry Carew and Alfonso de Salvio in edition published by Dover Publications Inc, New York, nd, 215
10 Newton, *Principia . . .* Translated by Andrew Motte and revised by Florian Cajori, University of California Press, Berkeley, 1934, 13
11 Newton, *Principia . . .* Translation as extract 10 above, 546
12 Halley, *A Synopsis of the Astronomy of Comets*. Translated from the original printed at Oxford. John Senex, London, 1705, 20

13 William Herschel, *Philosophical Transactions of the Royal Society*, [hereafter referred to as *Phil. Trans.*], vol 71, 1781, 492

14 Adams. From a memorandum now preserved in Adams' notebooks in St John's College, Cambridge, and quoted in Morton Grosser, *The Discovery of Neptune*, Harvard University Press, Harvard, 1962, 75

15 Leverrier, *Astronomische Nachtricten*, no 580, 1846, 53. Translated by Colin A. Ronan

16 John Herschel, *Athenæum*, London, 3 October 1946, 1019

CHAPTER II MEASURING THE STARS

17 Ptolemy, *Almagest*, Bk VII, section 5. Translation from same source as extract 2

18 Tycho Brahe, *Historia Coelestis*, Augsburg, 1666. Published posthumously. Translation by Colin A. Ronan

19 Flamsteed, *Historia Cœlestis Britannica*, London, 1725. Translation by Colin A. Ronan

20 Halley, *Phil. Trans.*, vol 30, 1717-19, 736

21 Bradley, *Phil. Trans.*, vol 35, 1729, 637

22 William Herschel, *Phil. Trans.*, vol 73, 1783, 247

23 William Herschel, *Phil. Trans.*, vol 93, 1803, 339

24 Bessel, *Monthly Notices of the Royal Astronomical Society* [hereafter *Mon. Not. R. astr. Soc.*], vol 4, 1838, 152

25 Bond, *Publications of the Astronomical Society of the Pacific*, vol 2, 1890, 300

26 Spencer Jones, *Mon. Not. R. astr. Soc.* vol 101, 1942, 356

CHAPTER III THE NATURE OF THE STARS

27 Aristotle, *De Caelo*, Bk II, 289ᵃ. Translation same source as extract 1

28 Tycho Brahe, *De Stella Nova*, signature *A*, Copenhagen, 1573. Translation by Colin A. Ronan

29 William Herschel, *Phil. Trans.*, vol, 85, 1795, 46

30 Kirchhoff, *Berlin Acad Bericht*, 1859, 662. Translated by H. Roscoe, *Spectrum Analysis*, London, 1870, 217

31 Huggins and Miller, *Phil. Trans.*, vol 154, 1864, 438

32 Letter from P. Janssen to the French Academy of Sciences, 26 October 1868, and quoted in Roscoe, *Spectrum Analysis*, London, 1870, 258. Translation by Colin A. Ronan

33 Cannon, *Journal of the Royal Astronomical Society of Canada*, vol 9, 1915, 203

34 Helmholtz, *Popular Scientific Lectures*, translated E. Atkinson, London, 1908, 178

35 Eddington, *Stars and Atoms*, Clarendon Press, Oxford, 1927, 99

36 Bethe, *Physical Review*, series 2, vol 55, 1939, 434

CHAPTER IV THE NEBULAE

37 Galileo, *Siderius Nuncius*, Venice, 1610. Translated by E. S. Carlos, Dawsons, London, nd, 42

38 Halley, *Phil. Trans.*, vol 29, 1714-16, 390

39 Messier *Connaissance des Temps pour* 1784, Paris, 1781, 232. Translated by K. Glyn-Jones in *Journal of the British Astronomical Association*, vol 79, 1969, 361

40 William Herschel, *Phil. Trans.*, vol 101, 1811, 269

41 Rosse, *Phil. Trans.*, vol 151, 1861, 681

42 Huggins, *Publications of Sir William Huggins's Observatory*, vol 1, 1899, 11

43 Huggins, *Phil. Trans.*, vol 158, 1868, 546

CHAPTER V   THE LAYOUT OF THE UNIVERSE

44 Aristotle, *De Caelo*, Bk II, 286[b]. Translation from same source as extract 1

45 Wright, *An Original Theory or New Hypothesis of the Universe*, London, 1750, 62

46 William Herschel, *Phil. Trans.*, vol 74, 1784, 437

47 Kapteyn, *Report of the British Association for the Advancement of Science*, Section A, London, 1905, 258

48 Leavitt, *Harvard Circular*, No 173, Harvard, 1912

49 Shapley, *Star Clusters*, Harvard Monograph No 2, McGraw Hill, New York, 1930, 171

50 Oort, *Bulletin of the Astronomical Institute of the Netherlands*, vol 3, 1927, 275

51 Reber, *Astrophysical Journal*, vol 91, 1940, 621

52 Hubble, Humason, *Astrophysical Journal*, vol 74, 1931, 43

53 Baade, Minkowski, *Astrophysical Journal*, vol 119, 1954, 206

54 Dewhirst, *Paris Symposium on Radio Astronomy*, ed R. Bracewell, Stanford University Press, Stanford, 1959, 507

55 Hewish, Bell, Pilkington, Scott, Collins, *Nature*, vol 217, 1968, 709

CHAPTER VI   BEGINNING AND END

56 *Kant's Cosmogony*, translated by W. Hastie, Glasgow, 1900, 73

57 Laplace, *System of the World*, translated by J. Pond, London, 1809, vol 2, 362

58 Jeans, *Universe Around Us*, Cambridge University Press, Cambridge, 1929, 232

59 Hoyle, *Quarterly Journal of the Royal Astronomical Society*, vol 1, 1960, 28

60 Eddington, *Space, Time and Gravitation*, Cambridge University Press, Cambridge, 1923, v

61 Hubble, *Proceedings of the National Academy of Sciences*, vol 15, 1929, 173

62 Lemaître, *Nature*, vol 127, 1931, 706

63 Gamow, *Nature*, vol 162, 1948, 680

64 Bondi, Gold, *Mon. Not. R. astr. Soc.*, vol 108, 1948, 252

65 Ryle, Clarke, *Mon. Not. R. astr. Soc.*, vol 122, 1961, 349

66 Hoyle, Narlikar, *Mon. Not. R. astr. Soc.*, vol 123, 1961, 133

67 Dicke, Peebles, Roll, Wilkinson, *Astrophysical Journal*, vol 142, 1965, 415

68 Greenstein, Schmidt, *Astrophysical Journal*, vol 140, 1964, 2

69 McCrea, *Mon. Not. R. astr. Soc.*, vol 128, 1964, 335

Acknowledgements

I should like to take this opportunity of thanking particularly my wife for supplying illustrations from her collection, the Ronan Picture Library, and Dr R. E. W. Maddison, Librarian of the Royal Astronomical Society, for his very ready help over source material.

C. A. R.

Illustrations:
Ronan Picture Library: 1, 9, 11, 22, 28, 30-3, 35-6, 38-42, 44, 49-51, 53-9, 62, 65, 67, 71, 74, 78, 81
Ronan Picture Library and Royal Astronomical Society: Frontispiece, 2-7, 10, 13, 16, 17, 23-7, 29, 37, 45-6, 48, 63-4, 66, 69
Royal Astronomical Society: 73
Ronan Picture Library and E. P. Goldschmidt & Co Ltd: 8, 15
Lowell Observatory: 12
NASA: 14, 20-1, 89
Hale Observatories: 18, 43, 47, 60-1, 70, 75, 77, 80, 83-4
Novosti Press Agency: 19
Royal Observatory, Edinburgh: 34
Lick Observatory: 52, 76, 79, 87-8
Harvard Observatory: 68
Astrophysical Journal: 72
Cornell University: 82
Bell Telephone Laboratories: 85
California Institute of Technology: 86